Chaos In Boxes
twisted adventures in music theory

Sean Alexander Luciw

print-on-demand edition

CHAOS IN BOXES
TWISTED ADVENTURES IN MUSIC THEORY

print-on-demand 1ˢᵗ edition
published by Produce Press
www.producepress.com

All rights reserved.
Copyright © 2011 Sean Alexander Luciw

No part of this book, covered by the copyright herein, may be reproduced or transmitted in any form or by any means – graphic, electronic or mechanical – without prior permission of the author, except for excerpts in a review. Any request for duplication of any part of this book shall be directed to the authour:
twistedmusictheory@gmail.com

Library and Archives Canada Cataloguing in Publication

Luciw, Sean, 1971-
Chaos in boxes : twisted adventures in music theory / Sean Alexander Luciw.

Includes bibliographical references.
ISBN 978-0-9784627-2-7

1. Music--Miscellanea. 2. Music theory.
1. Musique--Miscellanées. 2. Théorie musicale.
I. Title.

ML63.L938 2009   780.2   C2009-904821-3

Chaos In Boxes
twisted adventures in music theory

Table Of Contents

Section I: Introduction
    You Are Now Consuming My Brain   1
    A Really Cool Dream I Had   3

Section II: Infinitessimalitudinaciousness
    The Unattainability Of Infinity   5
    The Eye Of Babooshka   9
    The Golden Mean - It Ain't So Mean   13
    Meet You On The Dark Side Of The Mean   16
    Ceres - Master Of The Mirror Key   19
    Phi in Bode's Law?   22
    Skein And Shout   26
    Poetry Via Numerology   28
    Music Via Numerology   29

Section III: Mighty All-Pervading Power Chord
    Musical Interval As Distance   31
    Musical Interval As Ratio   36
    Perfect Fifth To The Power Of Twelve   39
    Temper, Temper!   42
    Quincunx And The Legend Of The Chromatic
      Transformation Matrix Starflower   45
    The Palindromic Di-Componential
      Chromatic Scissional Rotation Procedure   52
    The Golden Interval   54
    Golden Arpeggios   55
    The Times-Table Mirror   57
    The Modal Mirror (Bumblebees & Basketballs)   62
    Rock Stars   65

Section IV: Deep Light
    Light Equals Sound   69
    Sound Came First   76
    Has Anyone Seen The Music?   79

Section V: Exploring The Ancient Solfeggio Matrix

Introducing The Ancient Solfeggio Frequencies	83
Solfeggio Octaved Skeins	87
The Solfeggio Transposition Triangles	90
Solfeggio Transposition Triangle Nesting	93
Solfeggio Platonics	94
A Nasty Surprise	101
Massive Solfeggio	103
Solfeggio Intervals	105
The Quasi-Palindromic Triangular Solfeggio Sequence	110
Solfeggio Fretboard	113
Circular Graphs Seem To Reveal A Lot	114
Solfeggio Poetry Via Numerology	117

Section VI: Appendices

Appendix 272659.6521
    Converting Light Wavelength To Sound Frequency   119

Appendix 5
    Hertz Donut   120

Appendix A
    The Infamous Australian Treble Clef Dilemma   120

Appendix GGG
    A Symmetrical Joke   121

Appendix 100
    Golden Nice   121

Appendix Φ
    The Golden Ratio With Lots Of Decimal Places   122

Appendix .
    A Word On Decimal Places   123

Appendix 4
    Number-Boiling   124

Appendix 4
    The Circle Of Fourths   127

Appendix 4:44:44 04/04/04
    Eight Fours In A Row   128

Appendix 0.55555
    5wimming   130

Appendix 5
    F.O.S.F. (Found On Sidewalk Five)   131
Appendix 14
    F.O.R.F. (Found On Road Fourteen)   132
Appendix 142857
    1/7 As A Decimal   133
Appendix Φ
    The First 34 Fibonacci Numbers   135
Appendix EOOEOOEOOEOO...
    The Fibonacci Waltz   136
Appendix 21
    Fibonacci 21: A Musical Exploration   137
Appendix 21
    Bottle Cap   142
Appendix 6693
    Clock Pie   143
Appendix 6693
    Ancient Solfeggio Allows Time To Exist?   144
Appendix Φ
    Pythagorean Skein Reduction Of
    Successive Decimal Depths Of
    The Golden Ratio   145
Appendix 1497
    Palindromic Skein Sequence
    In The Squares   147
Appendix 364.5
    The Orbital Gnomon Star Power Chord   148
Appendix 365
    Gas Bill   150
Appendix ±
    The Margarine Era   151
Appendix $Z = Z^2 + C$
    Mandelbrot Doppelgänger   153
Appendix 6.4
    Koch Snowflake   154
Appendix 6.5
    Koch Tetrahedron   156
Appendix 6
    Shine On, You Crazy Tetrahedron   158
Appendix 8
    Allotropic Faces   159
Appendix 7
    The Two-Tone Tuning Fork   160

Appendix 1:2:3
    Mercury Tetractys   160
Appendix 11:22:33:44:99
    The Sunspot Dozen   161
Appendix 144
    12-Bar Blues   162
Appendix 420
    Re-Factoring The Calendar   164
Appendix 13
    The Fat Orphan   165
Appendix 33
    Solfeggio Factors   166
Appendix S
    Entropy   167
Appendix S-VHS
    Entropy (VHS Remix)   169

Section VII: Glossando

Section VIII: Bibliography

Section IX: Acknowledgments

Section I

INTRODUCTION

*Everything is made of music.
Sound is number, and the structural
framework of reality itself consists of
mathematics and musical harmony.
Various notions such as astronomy, art, anatomy,
philosophy, geometry, numerology and architecture
are not as separate as they might seem.
Hidden connections are revealed by unexpected keys.
Everything is a metaphor for everything else.
All is one, one is all.*

SECTION I: INTRODUCTION

CHAPTER Φ

You Are Now Consuming My Brain

Hello my name is Sean and this is my book about music and numbers and stuff. Although guitar is still my #1 favourite musical instrument to play after 20 years-plus of musical life, I like to branch out into other instruments, new styles of music, and other forms of expression such as painting or writing. It keeps things interesting, with an added bonus effect: experience in one medium enhances experience in other mediums. I'm often amazed at the overlaps that occur between presumedly separate aspects of reality. Some are easy to fathom, such as the similarity between a guitar and a bass guitar - they both have strings and frets. But... when I found out that planets Venus and Mars are singing a heavy metal power chord harmony out there in space, I was shocked, mystified and inspired.

Music: so much more than just a hobby, career choice, beer salesman or aphrodisiac! Everywhere you look, throughout the whole entire universe, music is at the heart of it all. Astronomy, biology, numerology, geometry, physics, art, architecture - musical concepts of rhythm and harmony are always popping up in unexpected places. It seems that music is everywhere, even in silence.

A short poem to celebrate the notion:

>a musical note
>equals
>a full moon
>equals
>a heart beat
>equals
>a light wave
>equals
>a silk hanky in the wind
>equals
>a sand crab skittering crabbily over the sand
>equals
>a dog shaking the water out of its fur
>(and the sand from its blanket)

In my various creative adventures (musical and otherwise), I've explored various polarities such as chaos vs. order, dark vs. light, opposite sides of the colour wheel, emotionless logic vs. logicless emotion, abstraction vs. structure, tonality vs. spectralism, the common mind vs. the individual mind... and various states of balance between. Music theory - the technical nuts & bolts of composition expressed through number and letter systems - has always captured part of my attention. It's a great asset to any musician, for it not only provides answers to technical problems, but hints at new mysteries for the composer's mind to unravel.

Every number has its own resonance. Nature's equations create shapes for the ears, mind, body and soul. Knowing this, and having access to the internet as well as used book stores and the library down the street, I've been on a bit of a research rampage for the last few years. While trying to absorb at least a humble smidgeon of the gargantuan body of knowledge that's already out there, I've been mostly concerned with contributing new theories and ponderances by weaving through my own explorations.

Chaos In Boxes represents some of my most interesting findings in this endless adventure. If you enjoy them at least half as much as I have, then the world is a happy place.

Sincerely,

Sean Alexander Luciw

CHAPTER Φ

A Really Cool Dream I Had

A few years ago (around my quarter-century mark or so), I had a kooky dream:

> On foot, I approach a house at the end of a cul-de-sac on the North Shore in my hometown, in the same neighbourhood where I went to Kindergarten in real waking life. A shady willow stands in the front yard, and the driveway is gravel. Up the concrete stairs and in through the front door, I enter into my living room studio where all of the high-precision musical gear I could ever want or use, and a huge set of speakers stand at the ready on a red and purple two-tone shag carpet. With the sun barely squeeking through the thick red curtains, there was a gentle darkness to the room.
> The overall visual effect was remininiscent of hovering above a rumbling volcano.
> As I begin to power up the machines, chuckles of delight sprinkle out at the thought of my assignment: I have been assigned to discover the medical secrets of sound, and create healing music. Specifically, I must find the resonant and anti-resonant frequencies of each of the umpteen organs within the human body, and any other type of body for that matter. By finding the proper frequency, I would be able to activate healing in certain body parts, or, conversely, destroy harmful toxins or malignancies by projecting the appropriate combination of sound frequencies towards the area in question.

A low C-sharp might clear a blocked artery, for example. I'm guessing that a J-flat minor seven teeth chord with a flat second-and-a-half would reverse tooth decay.

Oh yeah, guess who's the employer in this dream - NASA!

As it turns out, this sort of technology is already being put into action, and not just some crazy dream. Regardless of any prophetic value this dream may have had concerning my world or the world at large, I still found it inspiring.

Section II

INFINITESSIMALITUDINACIOUSNESS

Infinity is tough to fathom.
Really big stuff, really small stuff - it really could be a lot bigger
(or a lot smaller).
And no matter how much bigger it is
(whatever "it" is),
it still ain't <u>nuthin'</u> compared to infinity!
Every new answer seems to imply a thousand more questions.
Frustrating as this may be, it can also be an inspiration.
Metaphorical bridges of entertaining and mysterious
cuteness abound between the various
thought-realms of our world.
Prepare to meet the Golden Number and numerology,
two very interesting keys to the mystery.

CHAPTER Φ

The Unattainability Of Infinity

The following anecdotal cliché illustrates the unattainability of infinity:
"Are we there yet?"
"Nope."
"Are we there yet?"
"Nope."
"Are we there yet?"
"Nope."
"Are we there yet?"
"Nope."
"Are we there yet?"
"Nope."
"Are we there yet?"
"Nope."
"Are we there yet?"
"Nope."
"Are we there yet?"
"Nope."
"Are we there yet?"
"Nope."
"Are we there yet?"
"Nope."
"Are we there yet?"
"Nope."
"Are we there yet?"
"Nope."
"Are we there yet?"
"Nope."
"Are we there yet?"
"Nope."
"Are we there yet?"
"Nope."
"Are we there yet?"
"Nope."
"Are we there yet?"
"Nope."

"Are we there yet?"
"Nope."
"Are we there yet?"
"Nope."
"Are we there yet?"
"Nope."
"Are we there yet?"
"Nope."
"Are we there yet?"
"Nope."
"Are we there yet?"
"Nope."
"Are we there yet?"
"Nope."
"Are we there yet?"
"Nope."
"Are we there yet?"
"Nope."
"Are we there yet?"
"Nope."
"Are we there yet?"
"Nope."
"Are we there yet?"
"Nope."
"Are we there yet?"
"Nope."
"Are we there yet?"
"Nope."
"Are we there yet?"
"Nope."
"Are we there yet?"
"Nope."
"Are we there yet?"
"Nope."
"Are we there yet?"
"Nope."
"Are we there yet?"
"Nope."
"Are we there yet?"
"Nope."
"Are we there yet?"
"Nope."

SECTION II: INFINITESSIMALITUDINACIOUSNESS

"Are we there yet?"
"Nope."
"Are we there yet?"
"Nope."
"Are we there yet?"
"Nope."
"Are we there yet?"
"Nope."
"Are we there yet?"
"Nope."
"Are we there yet?"
"Nope."
"Are we there yet?"
"Nope."
"Are we there yet?"
"Nope."
"Are we there yet?"
"Nope."
"Are we there yet?"
"Nope."
"Are we there yet?"
"Nope."
"Are we there yet?"
"Nope."
"Are we there yet?"
"Nope."
"Are we there yet?"
"Nope."
"Are we there yet?"
"Nope."
"Are we there yet?"
"Nope."
"Are we there yet?"
"Nope."
"Are we there yet?"
"Nope."
"Are we there yet?"
"Nope."
"Are we there yet?"
"Nope."
"Are we there yet?"
"Nope."

"Are we there yet?"
"Nope."
"Are we there yet?"
"Nope."
"Are we there yet?"
"Nope."
"Are we there yet?"
"Nope."
"Are we there yet?"
"Nope."
"Are we there yet?"
"Nope."
"Are we there yet?"
"Nope."
"Are we there yet?"
"Nope."
"Are we there yet?"
"Nope."
"Are we there yet?"
"Nope."
"Are we there yet?"
"Nope."
"Are we there yet?"
"Nope."
"Are we there yet?"
"Nope."
"I gotta go pee."
"We're almost there."
"Are we there yet?"
"Nope."
"Are we there yet?"
"Nope."
"Are we there yet?"
"Nope."
"Are we there yet?"
"Nope."
"Are we there yet?"
"Nope."
"Are we there yet?"
"Nope."
...ad infinitum

SECTION II: INFINITESSIMALITUDINACIOUSNESS

CHAPTER Φ

The Eye Of Babooshka

What lurks within the bowels of this owl-shaped babooshka?

FIG. 4:   OWL-SHAPED BABOOSHKA FOUND ON SOME MISCELLANEOUS INSPIRED TREASURE HUNT

More babooshkas, of course...

FIG. 4:   MATRESHKI, N.T.S.

...but, what *else*? Something deeper, more cosmic... the secrets of *infinity,* perhaps?

Each babooshka contained within is progressively smaller than the last - sort of like looking into two mirrors which are facing each other. This can be metaphorically represented in the following number pattern, which I like to call the *babooshka fractions*:

1/2   1/4   1/8   1/16   1/32   1/64

The first fraction, 1/2, represents the outermost shell of the babooshka system. As the bottom half of each fraction gets larger by a factor of 2, each babooshka gets smaller in size by a musical octave.* On and on this procedure is repeated, revealing a smaller and smaller babooshka each time until no smaller babooshka could be created using current babooshka-construction technology. If this practical limitation of craftsmanship did not exist, the diminishing series of babooshka fractions would continue *forever and ever* into the depths of *infinitessimalitudinaciousness*.

This is where it gets interesting... if you add up all the babooshka fractions:

1/2 + 1/4 + 1/8 + 1/16 + 1/32 + 1/64 = 63/64
(only a sliver short of the fullness called *ONE*)

No matter how many more babooshka fractions you add to the series, the sum total will *never reach 1*.** A little further exploration to illustrate my point:

1/2 + 1/4 + 1/8 + 1/16 + 1/32 + 1/64 + 1/128 + 1/256 = 255/256
(closer but still no cigar)

In my perception, this procedure hints at the unattainability of infinity... But wait, there's more!

In Egyptian cosmology there was a character named Horus (real Egyptian name *Heru*), whose eye - purportedly plucked by his uncle Seth and subsequently re-installed by the gods - has been customarily represented thusly:

---

* Refer to Section III for discussion of the octave and other musical intervals. Rhythmic connotations also abound (whole notes, half-notes, quarter-notes, quavers, hemidemisemiquavers...)
** You may wish to look up *Zeno's dichotomy* sometime.

Not just a pretty face or a symbol of protection and power, the Eye of Horus was also given mathematical connotations, and was used as a teaching device in this regard. The various pieces of the picture represent pieces of the whole. Each subcomponent of the Eye of Horus refers to one of the aforementioned babooshka fractions:

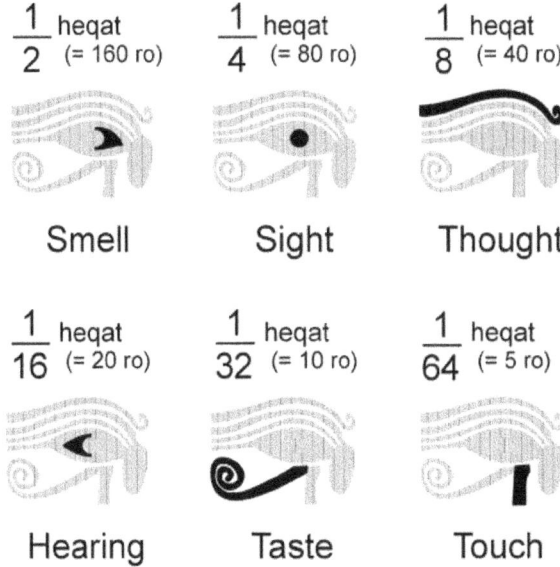

Anatomy of The Eye of Horus

As you can see, the Eye of Horus was not only used to teach math, but also to measure stuff (*ro* means *mouthful* and *heqat* means whatever heqat means), and its components also represented six senses. Since the fractions do not quite total one, I enjoy interpreting this last detail to mean that *all of our available senses could never give a complete picture of reality.*

The world is so big and I am so small. A lifetime is but a grain of sand on the endless cosmic beach of time!

Because of their common ground in the area of octave-based division, I perceive (and somewhat declare) that the Horus fractions and the babooshka fractions jointly allude to the *unattainability of infinity*.

CHAPTER Φ

The Golden Mean - It Ain't So Mean

Let's play a number game. We'll generate a famous series of numbers called the Fibonacci Sequence. Start with *zero* as the first number and *one* as the second number. To establish the third number, add the first two. What do you get? *One.* Add the second and third number to get the fourth number. What do you get? *Two.* The jist of the game: continue adding the last two numbers to create the next number in the series:

0 1 1 2 3 5 8 13 21 34 55 89 144 233 377
610 987 1597 2584 4181 6765 10946 17711
28657 46368 75025 121393   ...ad infinitum

Already an interesting game in itself which could provide almost anyone with limitless enjoyment,* this list of numbers goes way above the call of duty when it comes to unlocking the secrets of the universe. The simple list of numbers you see before you holds the primary active ingredient in many a secret formula from nature's cookbook.

By dividing any two side-by-side members of the Fibonacci Sequence (choosing a larger pair increases accuracy), a mystical dividend is unveiled. This secret agent, as with all secret agents, uses a few different aliases - the Golden Mean,** the Divine Proportion, the Golden Ratio, or simply the Golden Number. It is also known by its Greek name and letter symbol *phi*, which looks like this:

Φ

...the pronunskiation of which seems to be up for debate; is it *fye* or is it *fee?* I say fye.

Phi is partway between one-half and two-thirds, and just a little more than three-fifths: approximately 0.6180339...*** It's an irrational number**** - its decimals unfold forever without repeating.

---
* See Appendix 21, "Fibonacci 21: A Musical Exploration."
** In philosophy, *Golden Mean* describes a perfect balance between the two extremes of excess and deficiency. Just a tidbit of cross-disciplinary trivia I thought you might be interested in.
*** See Appendix Φ, "The Golden Ratio With Lots Of Decimal Places."
**** The word "number" is of questionable appropriateness, since it seems to imply that something is being *counted;* Phi does not count things - it describes a proportion or a ratio - a numerical *shape,* per se.

Phi's defining property, the quality that makes it so amazing, is reflected in the maxim *as above, so below*. In geometrical terms, phi is the perfect division of a line into two segments of very specific lengths, such that the smaller piece divided by the larger piece is the same ratio as the larger piece divided by the entire length:

$$A \bullet \text{————} B \bullet \text{——} C \bullet$$

$$BC / AB = AB / AC = \Phi$$

This division can be reiterated in either direction to create both shorter and longer pieces of line segment, all of which conform precisely to the exact same symetry. No matter to what degree one chooses to extrapolate or interpolate ripititions of the Golden Mean, the self-reflective property is always maintained.

As above, so below.

There is theoretically no limit to the degree of expansion and contraction with this process. Zoom in further and further and further and you will find the same shape as you started with. Zoom out and out and out and the picture will still be the same. This property is commonly known as *self-similarity*. The perfect dynamic balance of phi is a never-ending vortex.

Aurea mediocritas - up supra ut infra. Latin - a language I do not yet remember how to speak.

According to popular convention, Phi with a capital *P* specifies 1.618..., and with lower case *p* implies 0.618... Either way you slice it, it's all the same bag of chips. Zero point six one eight zero three three..., 1.618033..., 2.618033..., 3.618033..., etcetera... they're all outward echoes of the Golden Ratio.

The following formulae illustrate Phi's ouroboric tendencies:

$$\text{Phi} \times \text{phi} = 1$$
$$\text{Phi} / \text{phi} = \text{Phi} + 1$$
$$\text{Phi}^2 = \text{Phi} + 1$$
$$\text{phi} / \text{Phi} = 1 - \text{phi}$$
$$\text{Phi} - \text{phi} = 1$$
$$\text{phi}^2 = 1 - \text{phi}$$

There are several methods of "finding" this magic number. As already mentioned, the Fibonacci Sequence is one path to the Golden Mean. Another convenient place you'll find Phi is within the perfect pentagram:

(see diagram next page)

SECTION II: INFINITESSIMALITUDINACIOUSNESS

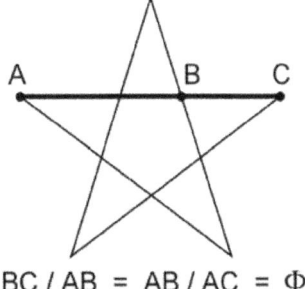

$BC / AB = AB / AC = \Phi$

Phi and the pentagram are found naturally in the motions of celestial bodies (Venus draws a pentagram around the Earth every thirteen years, and nested groups of pentagons describe various orbital shapes in our Solar System), and all sorts of other intriguing areas of our universe - the stem angles of some plants (I have seen these shapes with my very own eyes - it's true, the vegetation really knows its math), the spirals generated by pineapples and sunflowers (or, does the spiral generate the fruit?), seashell shapes (the Chambered Nautilus in particular), human anatomy (the belly-button's height from the floor compared to total height is only one example of many), the designs on a butterfly's wing... the list goes on. The ancientness of the Big Bang divided by the age of the Sun is also Phi.*

I shit you not!

Some humans believe that the Golden Proportion is aesthetically preferable over other measurements; many artists, musicians and architects throughout history have used the Golden Number in their output. The ancient Egyptians built a Phi spiral into the layout of the pyramids and the Sphinx at Giza. Stradivarius used Phi to determine the f-hole placement in his famous violin designs. Béla Bartók and Erik Satie are two famous composers who supposedly used Phi to determine factors such as rhythmic placement of significant motifs, although some of this is under debate. Photographers often organize their compositions in sections of three, which could be interpreted as an approximation of Phi. Credit cards are also in golden proportion - how poetic!

All of this amazingness from a simple number game.

Phi'll be back...

---

* Refer to Appendix 364.5 for a cross-disciplinary correlation involving Phi, of my own discovery.

CHAPTER Φ

Meet You On The Dark Side Of The Mean

Based on my establishment of a sequence for progression through the Solfeggio matrix (an ancient musical scale described in Section V forthcoming), I recently and suddenly felt inspired (last night, after midnight) towards an attempt to create a sort of "reverse-engineered" Fibonacci Sequence beginning with the number pair of 71 and 44, and devolving backwards. Sorry to wreck the surprise,* but seventy-one fourty-fourths is the fraction which occurs in the Solfeggio as the intervals Fa-over-Ut and La-over-Mi; and, even though the number pair does not appear in the classic Fibonacci Sequence, they do deliver a quotient of 1.6136363636363636..., which is quite close to the Golden Ratio. So it only made sense to try and extrapolate a sequence...

Here's what happened, starting with 71 and 44 on the right-hand side and devolving to the left towards zero by subtraction:

... 73 -45 28 -17 11 -6 5 -1 4 3 7 10 17 27 44 71

Seventy-one minus fourty-four is twenty-seven; fourty-four minus twenty-seven is seventeen; twenty-seven minus seventeen is ten; seventeen minus ten is seven; ten minus seven is three; seven minus three is four... Strange: everything seemed to be going smoothly until it was 7's turn to break down - four and three are in reversed order! This is something I did not expect. And, whereas the "real" Fibonacci series is sprouted from the "binary seed" of zero and one, the above sequence does not even contain zero *or* one.**

Weird.

---

* See p. 107.
** I'm sort of disappointed to announce that I've since discovered that *any* two numbers will automatically result in a sequence which yields the Golden Mean upon division of any two of its more highly-evolved components, thereby somewhat trivializing the Fibonacci Sequence's primordial pair of 0 and 1 that I'd always thought was such a big deal. Drats!

(eg. the un-Fibonacci-ish pair of 4 and 19: 4 19 23 42 65 107 172 279 451 730 1181 1911 3092 5003 8095 13098 21193 34291 55484 89775 145259... 145259/89775=1.618033...)

This seems to have an interesting implication in the context of cosmological evolution: no matter what the origin of the Universe was like, *something* as intricate and awe-inspiring as the current situation would have been *inevitable;* and if high order is the default destiny of *all* chaotic beginnings, the human race itself is more like a by-product of time than some miraculous creation. I wonder what the gods have to say about *that?*

SECTION II: INFINITESSIMALITUDINACIOUSNESS 17

I noticed something else: modulating polarity. To the left of where zero ought to be, the components of the series do a little flip-flop routine with their plus and minus signs. (Four minus negative one is five; negative one minus five is negative six; five minus negative six is eleven, etcetera...)

Which leads to the question: does this flip-flop modulation phenomenon also occur with the "real" Fibonacci Sequence? Maybe we just never noticed it 'cause nobody was looking in that direction.

Extrapolating to the left of zero:

... -21 13 -8 5 -3 2 -1 1 0 1 1 2 3 5 8 13 21 ...

Yes, it does! Wow, what an amazing smell I've just discovered. "Scent Of A Modulating Polarity," out now on VHS and Beta. Not only do the plus and minus dance back and forth, but the left side of zero is a mirror image of the right - both sides include the same numerals evolving in opposite directions away from zero. Just like sine waves, the breath, the Big Bang Theory and other assorted oscillation-oriented metaphorical connections which the reader is free to discover.

Next question: as is common knowledge, the ratio of successive Fibonacci terms approaches the value of Phi throughout its evolution away from zero. So, what happens when we use the newly-extrapolated portion of the series which is to the *left* of zero?

Thirteen divided by minus eight is approximately negative Phi.

Minus twenty-one divided by thirteen is also approximately negative Phi.

Well, that makes it simple; the negative side of the Fibonacci Sequence generates *negative Phi* and the positive side generates *positive Phi*. As one might expect, I suppose. No big deal, but I guess it was still worth figuring out. Math students probably know all this stuff already. I'm just a guitar player in mathematician's clothing.

Useless information - is it a modern epidemic, or a contradiction in terms? Discuss amongst yourselves.

Every second term in the positive Fibonacci Series has a negative twin on the minus side of zero. And so I ponder - is there any special significance or property of those numbers in the sequence which acquire a minus sign versus those which don't? This leads me to separate the Fibonacci Series into two *subcomponents* (this reminds me of the Palindromic Di-Componential Chromatic Scissional Rotation Procedure forthcoming in Section III...):

1 3 8 21 55 144 377 987 2584 6765 17711...
(modulating subcomponent A)

and:

1 2 5 13 34 89 233 610 1597 4181 10946...
(non-modulating subcomponent B)

Which leads to the next question - what is the ratio of adjacent terms of each subcomponent series?

17711 / 6765 = 2.61803399...
(modulating subcomponent A)

and:

10946 / 4181 = 2.61803396...
(non-modulating subcomponent B)

...which are both basically Phi.
This still leaves the mystery of the modulating polarity unanswered; what is the significance of these subcomponent serieses?
Cerises?
Ceres?

SECTION II: INFINITESSIMALITUDINACIOUSNESS   19

CHAPTER Φ

Ceres - Master Of The Mirror Key

Ceres, the biggest asteroid in the belt, has a pretty amazing function in the mathematical symetry of the Solar System. It was discovered by Alex B. Geddes in the late twentieth century and popularised by John Martineau - a simple mathematical phenomenon in which Ceres acts a sort of multiplication mirror between the four inner planets and the four outer gas giants.* It goes a little something like this:

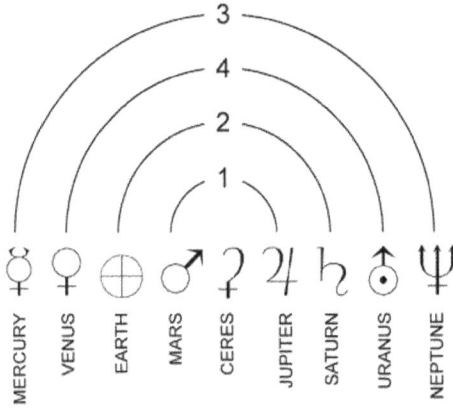

CERES MULTIPLICATION MIRROR
(incarnation #1)

In the above diagram, the planets are arranged from left to right in increasing distance from the Sun. Pluto doesn't get to play in this reindeer game. Sorry. Maybe Indigo's free this weekend...**

Multiply the *orbital radii* of the two planets referred to as pair # 1 (the two closest to Ceres - Mars and Jupiter).

Next, multiply this number by 1.206. The result is equal to the product (multiplication) of orbital radii of pair #2 (Earth and Saturn), which is the next furthest from Ceres.

---

* See Appendix GGG.
** See p. 71.

Here are the scribbles:

planet pair #1: 227.94$_{Mars}$ x 778.57$_{Jupiter}$ x 1.206 = 214025.5
planet pair #2: 149.60$_{Earth}$ x 1433.5$_{Saturn}$ = 214451.6
(all distances are x 10^6 km)

According to the Ceres mirror phenomenon, these two results should theoretically be the same:

214025.5 / 214451.6 = 0.99801307148093089

99.8 % - that seems fairly accurate.

The amazing thing about the system: the other planets exhibit the same formulaic coherence, mirrored about Ceres!* Here's the complete list of orbital radius parallels:

planet pair #1 x 1.206 = planet pair #2
planet pair #2 x 1.208 = planet pair #3
planet pair #3 x 1.204 = planet pair #4

That is,

1.206 x Mars x Jupiter = Earth x Saturn
1.208 x Earth x Saturn = Mercury x Neptune
1.204 x Mercury x Neptune = Venus x Uranus

Great stuff! Incredible, endless surprise patterns in the structure of our universe! Hmmm... where did this number *one-point-two* come from, anyways? What does it mean?** A new universal constant, perhaps?

In the words of Captain Manzini from *The Garbage Pail Kids Movie,* "I have a feeling the answer is a musical one."

---

*However, the rest of the system is not necessarily arranged how one might expect. Specifically, notice how, starting from the center of yesterpage's diagram, the planet pairs are labeled in the non-sequential order 1 2 4 3. Is Mother Nature dyslexic?

** Maybe the formula $\pi = 1.2 \times \Phi^2$ has something to do with it!

Another incarnation of the Ceres mirror phenomenon is as follows:

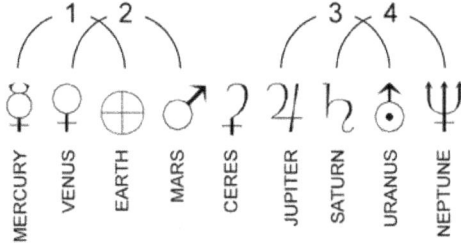

CERES MULTIPLICATION MIRROR
(incarnation #2)

This variation has its own set of formulae, with a different constant:

planet pair #1 x 2.872 = planet pair #2
planet pair #3 x 2.876 = planet pair #4

Or, in other words,

2.872 x Mercury x Earth = Venus x Mars
2.876 x Jupiter x Uranus = Saturn x Neptune

Awesome!* The planets seem to have been spun unpredictably from chaotic clouds of dust, and yet they exhibit amazing symetry and beauty in their design. Apparently those heavenly bodies are no slouch in the brains department, either!

---

* Also, if you multiply the orbital radii of planets Venus, Mars, Jupiter, and Uranus, the result is the same as multiplying those of Mercury, Earth, Saturn and Neptune.

CHAPTER Φ

Phi In Bode's Law?

There exists a simple and amazing number series which describes the orbital radii of the planets in our Solar System. The complicated history of its discovery can be found elsewhere, but I can at least tell you that it was defined in the late 1700s and has been popularly referred to as Bode's Law ever since. Other names were involved in its discovery and description, but in the interest of sheer conformity I'll simply refer to it as Bode's Law from here on in.

In any case, the formula I am about to describe *did* make one very big ripple in the perceptual pool of stargazing culture: it predicted a planet where there was none. A few years later, on New Year's Day of 1801, Giuseppe Piazzi discovered Ceres, the biggest asteroid in the asteroid belt, in the exact location predicted by Bode's Law. Everybody was amazed.*

To generate the number series, start with 0 as the first number and 3 as the second number. Repeatedly double 3 to generate the following:

0 3 6 12 24 48 96 192 384

The next step is to add 4 to each of these numbers, and you end up with the official Bode's Law number series:

4 7 10 16 28 52 100 196 388

The amazing thing about this bunch of numbers is that each number closely approximates the orbital radius of a planet, with Earth at 10 "Bode units" of 14.9725×10^6 km per unit:

Planet	Bode #	Bode-Predicted Orbit (*10^6km)	Actual Mean Orbit (*10^6km)	Bode's Law Error
Mercury	4	59.89	57.91	+ 3.42%
Venus	7	104.81	108.20	− 3.23%
Earth	10	149.73	149.60	+ 0.086%
Mars	16	239.56	227.94	+ 5.10%
Ceres	28	419.23	413.94	+ 1.28%
Jupiter	52	778.57	778.57	0
Saturn	100	1497.25	1433.50	+ 4.45%
Uranus	196	2934.61	2870.99	+ 2.22%
Neptune	------	------	4504.30	------
Pluto	388	5809.33	5913.50	−1.76%

---

* If they only knew what Mars Sector Six would eventually reveal to Dr. George King about Maldek, they'd have been even more amazed.

SECTION II: INFINITESSIMALITUDINACIOUSNESS

To establish the exact value of one Bode unit, I used Jupiter as the zero-point tuning reference because it's the hugest planet, which in my opinion makes it the most steadfast and reliable, unlikely to be swayed by the breaking winds of space. I also feel compelled to mention that, during the writing of this book, Jupiter's Great Red Spot has acquired a smaller counterpart, and they are headed for a collision course! I predict they will simply meld into one greater red spot. I also predict that there are more on the way. Perhaps the red will spread and Jupiter will literally become "the other red planet."

The day I learned of this marvelous numerical phenomenon known as Bode's Law, I noticed a suspicious anomaly: who's missing?

Neptune!

Neptune doesn't appear in Bode's series of numbers. Why not? I have no idea! Curiously enough, though, some brief reading on the topic of astrology has told me that Neptune is "...the planet of unlimited possibilities and things we cannot touch or see..." and something to do with *camouflage*. How appropriate. I sure cannot see Neptune in the Bode list. And, in the famous symphony "The Planets" by Gustav Holst, Neptune is called *The Mystic*.

Coincidence, or **?**

Anyway, enough astrology already and back to astronomy. I mean, uhhhh... *mathematics*. Or, more specifically, geometry. Or whatever this is. Could somebody please tell me what category we're in now? I get so confused jumping back and forth between all these boxes. It's all really just one big sandbox, isn't it?

All of these -ologies and -onomies and what-have-you's...

I just happened to have a lot of the Golden Ratio on my mind on the day that I first became aware of Bode's Law. So, my first hunch and hope was that Neptune's missing mystery orbit would be somehow connected to Phi. Wishful thinking perhaps, but nevertheless I ventured forth to decode Neptune's mysterious absence from Bode's Law.

This is as simple as dividing Neptune's actual orbit by the size of a Bode Unit to determine its would-be Bode number, and then dividing the result by Uranus' and/or Pluto's Bode number and comparing it to Phi.

However, this also seemed like the perfect opportunity to have some fun putting the Ceres mirrors into action, so I chose to mix those into the procedure as well.

Plugging the appropriate Bode numbers into the first Ceres multiplication mirror, I solved for Neptune:

1.204 × Mercury × Neptune = Venus × Uranus
1.204 × 4_B × $Neptune_B$ = 7_B × 196_B
$Neptune_B$ = 284.88_B

and

1.208 × Earth × Saturn = Mercury × Neptune
1.208 × 10_B × 100_B = 4_B × $Neptune_B$
$Neptune_B$ = 302_B

Using the second Ceres multiplication mirror, I solved for Neptune again:

2.876 × Jupiter × Uranus = Saturn × Neptune
2.876 × 52_B × 196_B = 100_B × $Neptune_B$
$Neptune_B$ = 293.122_B

Next, I used actual planetary data to convert the orbital radius of Neptune to its Bode equivalent:

4504.3 × 10^6km / 14.9725 × 10^6km per unit = 300.838_B

These four different values all hover around about 301. This would be Neptune's place in the Bode series if it appeared there.

Here it comes - the moment of truth... will Neptune magickally conjure the Golden Number?

Subtracting 4 to get back to the initial "doubling" stage of Bode's Law, we get 297. According to my wishful intuition-slash-ambition, Neptune's 297 divided by the 192 of Uranus would equal the Golden Number.

So... *is it?*

(in a monster-truck commentator's voice: Two-hundred-and-ninety-seven divided by 192 equals...)
(drum roll, please...)

297_B / 192_B = 1.546875

Not exactly Phi, but sorta close. Hmmmmm... *How* close?

1.546875 / 1.618033 = 95.6022%

Actually, that's pretty accurate after all; in fact it's only low by 4.39%. Four-point-three-nine percent deviation might seem a bit sloppy,

SECTION II: INFINITESSIMALITUDINACIOUSNESS

except when compared to the other tolerances* which have already been accepted as part of Bode's Law, which has margins of more than 5 percent. These sorts of errors are partly reflective of orbital fluctuations caused by gravitational interactions between celestial bodies. Sort of like a guitar that goes a little out of tune in the presence of massive bass frequencies created by million-watt P.A. systems or because of ambient temperature changes. Or something like that.

In summary: my wish was granted after all - my hunch was correct. I solved the mystery. Lurking deep within Bode's Law, elusive Neptune holds the key to the Golden Mean; Neptune's "orbital seed index position" (for lack of a better term) in the Bode series relates to that of Uranus by a factor of Phi.

Eureka! One of my most important discoveries *ever.* You may bow to the king, but only if you bring gifts. Just kidding. You paid your twenty bucks or whatever for this here book; the information it contains is included in the price.

Another observation: the repeated doubling of 3 (in the first step of the Bode procedure) implies *ascending octaves* in a musical context. I'm not sure how the addition of 4 to each number (step 2) fits into the metaphor.

Another observation: if you re-arrange the letters of *mystery,* you get *symetry.*

---

* See Appendix ±, "The Margarine Era," for some discussion on margin of error.

CHAPTER Φ

Skein And Shout

Numerology is the practice of combining numbers and letters into one system. Letters and words can then be interpreted according to the sensibilities of number systems. Each number or letter carries its own unique resonant "flavour" of energy.

The letters of a word can be interpreted individually, or added together for a combined interpretation. If a number has more than one digit (eg. 23), then adding them together would reduce them to a single digit (eg. 23 becomes 5, because 2+3=5). I've seen this process referred to by the names "digit summing," "modulus 9" and "skein reduction."

According to the online dictionaries, "skein" means *a loosely coiled length of yarn or thread wound on a reel*, or *something suggesting the twists or coils of a skein.* The fully-reduced single digit equivalent of a number is sometimes called the Pythagorean Skein, named after a somewhat famous dead numbers guy. Reducing multi-digit numbers this way has the effect of creating a spiral coil of numbers which cycles from 1 thru 9, then wraps around back to 1 and up thru 9, over and over again:

Pythagorean Skein →

1	2	3	4	5	6	7	8	9
1	2	3	4	5	6	7	8	9
10	11	12	13	14	15	16	17	18
19	20	21	22	23	24	25	26	27
28	29	30	31	32	33	34	35	36
37	38	39	40	41	42	43	44	45

In the above table, any number in a column reduces to the number at the top of the column. We are witnessing mod 9 in action...

Consistent with the dictionary's suggestion, there is definitely a winding, twisting property to this numerical sequence. The grid shown above would wrap perfectly around a cylindrical base. If it were wrapped at a slight angle, the number 9 could be made to match up with the number 10, and 18 to 19, etcetera, and the spiral vibe would be more thoroughly manifested. If skeins 3, 6, and 9 were then coloured red with the others left blank white, a candy cane sort of phenomenon would result.

Sweet!

SECTION II: INFINITESSIMALITUDINACIOUSNESS

Substituting the alphabet for 1 through 26 on the previous page, we get the standard numerology grid:

Standard Numerology Grid								
1	2	3	4	5	6	7	8	9
A	B	C	D	E	F	G	H	I
J	K	L	M	N	O	P	Q	R
S	T	U	V	W	X	Y	Z	

And, in spiral form:

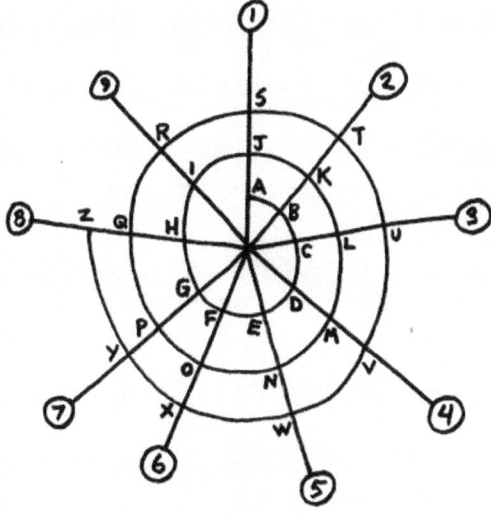

Hencely, the word "gobbledeygook" transforms into 7 6 2 2 3 5 4 5 7 7 6 6 2.

Adding these together for an overall resonance assessment, we find the number 8 (7+6+2+2+3+5+4+5+7+7+6+6+2=62, 6+2=8).

Also: more occurrences of a particular number accentuate the resonance of that number for the entire structure; "gobbledeygook" has an equal balance of 6, 7 and 2 leading the helm, with a lesser degree of 5, 4 and 3, and a complete absence of 1 and 9. Notice that, although the digit sum is 8, there is no 8 in the series.

Stay tuned for more fun with numerology-type stuff.

CHAPTER Φ

Poetry Via Numerology

Objective:
Compose a piece of poetry using a method based on the standard modulus-9 numerology grid:

Standard Numerology Grid								
1	2	3	4	5	6	7	8	9
A	B	C	D	E	F	G	H	I
J	K	L	M	N	O	P	Q	R
S	T	U	V	W	X	Y	Z	

Procedure:
1) Choose a few numbers from 1 thru 9:

    1 7 8

2) Look at the grid to find the matching letters:

    A J S G P Y H Q Z

3) Create a list of words using only those letters:

    JAH SAG GAP PAY SAY
    JAZZ GAS HAPPY   SAPPY
    YAPPY HQ GYPSY
                    etcetera...

4) Arrange the words into a literary masterpiece, such as the following screenplay:

> *Ah, say, Zappy...*
> *as jazzy gaps pay Happy,*
> *a gypsy sags?*
> *Pass gas as Jay gasps a sappy pap.*
> *Yap, yap, yap!*

5) Mission accomplished!

SECTION II: INFINITESSIMALITUDINACIOUSNESS   29

CHAPTER Φ

Music Via Numerology

Objective:
Compose a piece of music using a method based on the numerology system.

Procedure:
1) Choose a group of notes, and number them:

   A-sharp = 1   C-sharp = 2   D-sharp = 3
   F-sharp = 4   G-sharp = 5

2) Establish a letter-to-number grid to match the total number of notes chosen in step 1:

1	2	3	4	5
A	B	C	D	E
F	G	H	I	J
K	L	M	N	O
P	Q	R	S	T
U	V	W	X	Y
Z				

3) Choose a word and use the grid from step 2 to convert the word into a sequence of numbers, and then into musical notes:
   (eg. the word *arbitrary*)
   a=1=A#, r=3=D#, b=2=C#, i=4=F#,
   t=5=G#, r=3=D#, a=1=A#, r=3=D#, y=5=G#

4) Arrange the results from step 3 onto a clef and create music therefrom:

5) Mission accomplished!

Section III

MIGHTY ALL-PERVADING POWER CHORD

*By looking at musical structures through
the eyes of numbers and shapes,
some really amazing symmetries are revealed.*

CHAPTER Φ

Musical Interval As Distance

What is a musical *interval,* you ask?

Well, maybe you didn't ask, but you'll need to know a bit about intervals in order to understand the discussions of the long-lost ancient Solfeggio scale coming up in Section V.

An interval is basically the distance between two notes.

On the piano, for example, the distance between a C note and the next C note to the right (or left) is called one *octave;* on a guitar string, the octave is a distance of 12 frets.

In the traditional Chromatic Scale which underlies most popular music in the Western world, the octave is divided into 12 *semitones.* (On the piano, keys next to each other are a semitone apart.) This creates 12 distinct possibilities for interval distances from a single root. The total would be 13, if one were to count both unison and octave. However, the octave and the unison are functionally the same note; counting both as unique entities would be somewhat redundant. Therefore the Chromatic Scale should be treated as a 12-note system.*

The semitone, also known as a *half-step,* ** is the smallest available interval on standard Western instruments such as the guitar and piano; all other intervals are multiples of a semitone. The semitone may also be called a *major seventh* or a *minor second,* depending on the musical context, such as whether one is measuring upwards or downwards, or which scale tone you're measuring from. It can be a little confusing so, instead of opening that can of worms, I won't. Then again, I guess I kinda already did. To keep things consistent and sensible, I'll try to be consistent throughout the book and make sure I measure all intervals *upwards* from a single static point. In compositional terms this start point is called the *tonal center, tonic,* or *root* of the scale. In the key of C major, for example, C is the root note. In the key of A minor, A is the root.

---

* Have you ever contemplated the origin and meaning of the term *baker's dozen?* I asked a baker and he said that bakers often throw an extra bun into the dozen *just in case* - a sort of safety factor. Overabundance being more satisfactory to a paying customer than underabundance, I presume.

Safety in numbers - a bonus donut! And, since we're on the topic of the 12-13 binary, did you know that the planet Venus spins on her own axis 12 times in the same period of time it takes her to orbit the Sun 13 times?

See also Appendix 11:22:33:44:99, "The Sunspot Dozen" and Appendix 13, "The Fat Orphan."

** A *whole-step* equals two half-steps.

Here is a list of all 12 chromatic intervals and abbreviations:

Chromatic Intervals
(Distance in Semitones)

Interval Name	Symbol	DISTANCE (semitones)
Root (Unison)	R	0
Minor Second	m2	1
Major Second	M2	2
Minor Third	m3	3
Major Third	M3	4
Perfect Fourth	P4	5
Tritone	♯4 or ♭5	6
Perfect Fifth	P5	7
Minor Sixth	m6	8
Major Sixth	M6	9
Minor Seventh	m7	10
Major Seventh	M7	11
Octave	8ve	12

Each interval has its own sound, its own little mood. Labels such as "happy" or "sad" may be applied to the emotion of each interval, but the imprecision of language is equal to the subjectiveness and contextuality of music - and perception - itself.

It's the same with colours - I can say something is "blue" or, if I wanted to strive for accuracy, I could say, "a deep, melancholy cobalt blue" but there is no guarantee that you and I would imagine the same colour, or that it would look the same to both of us inside our brains if we saw it right in front of us.

There is a universe behind each face.

You could even say there is an interpersonal dynamic between notes. In music, a series of interactions between intervals causes shifts in tension (potential) and release (creation, resolution) and creates moving sculptures of colour and energy in the form of melodies and chords, brought to light through premeditated composition, rehearsed performance, spontaneous improvisation, auditory imagination, or some combination thereof.

Music is a great metaphor for all sorts of things in our universe - and, it's there for all to see... Nature really trips me out sometimes.

Maybe it's one of those need-to-see-it-to-believe-it kind of deals. Or, maybe it's a need-to-believe-it-to-see-it kind of thing.

This? That?

SECTION III: MIGHTY ALL-PERVADING POWER CHORD

ONE	THE OTHER
ONE	THE OTHER
ONE	THE OTHER
ONE	THE OTHER
ONE	THE OTHER
ONE	THE OTHER

Six of one, half-dozen of the other.

One way of memorizing the sounds of the musical intervals (and the scales that are created from them) is to find easy-to-remember-and-hear-in-your-head songs which use that interval. For example, the first two notes of the main riff to Black Sabbath's "Iron Man" are what I've always used to hear the sound of a minor third. "O Canada" also starts with an upward movement of a minor third; however, the first note of "O Canada" is actually the third of the scale, and not the root; the root of the scale does not appear until note #3... If you're as confused as I am then I've proven my point I was mentioning earlier about the can of worms - "Iron Man" is simpler to deal with because it *starts with the root* of its key. As another example, "Rock Around The Clock" by Bill Haley and the Comets is what I "sonically visualize" to remember what the dominant seventh arpeggio, and therefore the Mixolydian scale to some extent, sounds like. These sorts of reminders are helpful if I'm trying to write a melody on paper, or onto my memory, without having a musical instrument at hand to tell me which particular note names I'm really dealing with.

This skill of perception - identifying the intervallic makeup of melodies and chords - is commonly known as "relative pitch." Very enjoyable and enabling. However, there also exists "perfect pitch" - the ability to *name* a single pitch upon hearing it, without having access to a musical instrument to help determine which note is being sounded. Perfect pitch is sort of like having a flawless 1.000 average at piñata-batting. Perfect pitch is sometimes described as colour hearing, but you can definitely have colourful sound experience regardless of note-naming abilities. The notion of colour hearing also brings *synesthesia* (see Section IV) to mind.

You may have noticed from the previous chart that there are special names for all 12 intervals. Notice that there are two kinds of seconds, two kinds of thirds, two kinds of sixths, and two kinds of sevenths - a major and minor version of each, and the names only go up to "seventh." This is because the 12 names are geared to accomodate the high-variety world of 7-note scales. For example, the second step would either be called the *minor* second or the *major* second, depending on its distance from the root. There are two expressive options available there. For typical scale construction, you would not encounter both minor *and* major seconds simultaneously; the same goes for thirds, sixths, and sevenths - only one "type" of each scale "number" is found.*

The typical "Major Scale" which every music student has probably practiced and any person would probably recognize, is a perfectly straightforward example - it contains only major and perfect intervals: R, M2, M3, P4, P5, M6, and M7.

And the octave, of course. The octave is pretty much a foregone conclusion, and functions as a signpost on a moebius strip, where the scale is spelled the same way above the octave as below... extending in both directions (up and down), forever and ever beyond the ranges of our hearing and into the realms of light, heat, beats, and beyond. The Major Scale is the same arrangement of tones that one would stereotypically sing for a warm-up at the beginning of choir class while the teacher plays along on piano. The singing of these tones, along with their Latin names Do, Re, Mi, Fa, Sol, La, and Ti, is sometimes called the "Solfège."

The Solfège is not to be confused with the ancient "Solfeggio" frequencies, which are vaguely similar, yet completely different in all of the important numerical details. Stay tuned for an exhilarating examination of the ancient Solfeggio, coming up in Section V.

Anyways... even though only one type of each interval name typically appears in any diatonic scale, that is not to say that combinations of minor and major intervals would not, or should not, occur in real music - they often do! There's more to music than conformity to scales. For example, cycling chromatically upwards from the perfect fifth, through the minor sixth and then major sixth** and back down again is a very common and effective compositional phenomenon. "Kashmir" by Led Zeppelin, "The Call Of Ktulu" by Metallica, "Is There Anybody Out There?" by Pink Floyd, and the harmonies in the *James Bond* theme are all good examples which use this sort of "chromatic rollercoaster" of mixed minor and major intervals.

Heck, there are some tunes which use *every* note possible, all in the same song. "Flight Of The Bumblebee" is one such thing. If you want to play all 12 notes *at the same time*, simply lean on the piano keys (black *and* white) with your arms. Or, if the specifics of your limbs somehow prohibit this technique, it may be simpler to drop the piano from a few feet above the floor, thusly sounding *all 88 notes* at once. However, the tuning stability of the piano may be adversely

---

* Caveat: every rule in music is meant to be broken. Rules are merely convenient quantizations and useage guides for pre-existing understandings.
** ...and sometimes all the way up to minor seventh...

affected by this "chroma-drop" technique.

It's worth mentioning here that the 12-note Chromatic Scale is by no means the only possible way to divide the octave. The "quarter-tone" system divides the octave-pie into 24 equal pieces. Just imagine the beautiful variety of chords and melodies that could emerge from a system twice as refined as the standard 88-key piano! Rumour has it the 17-piece octave pie is a real sweet-sounding one - almost as nice as the 31. Twenty-two, 19... did you know that the touch-tone phone system is a 14-tone-per-octave division?

What's stopping anyone from inventing their own scale systems? Nothing. Go ahead, I won't stop you. Keep in mind here that an octave divided into *equal* pieces - five equal pieces in the case of Javanese gamelan music, for example - is not the same as a chromatically-based 5-tone pentatonic scale, whose tones are *unevenly* spaced (and selected from the 12-note Chromatic Scale).

From yet another perspective, one could even argue that any attempt whatsoever at subdividing the naturally continuous spectrum of tones into a grid-like, numerically-confined system amounts to a falsity, a bastardization and disintegration of nature's gradients rather than an improvement... yet another vain display of humanity's weird obsession with order.

Interesting purist mindset... useful as one side of a Yin-Yang, at least. I often find non-tonal music - I call it "ambient" - the most versatile to listen to, and the most natural to create in the recording studio - moreso than guitar music, even. In non-tonal mindset, any colour of sound can mix with any other, totally at the artist's discretion without boundary. This is a very different experience from the conformist extreme embodied in scale systems.

I figure a combination of both extremes makes for the most colourful music - a full spectrum of dynamic tensions and balances.

CHAPTER Φ

Musical Interval As Ratio

An important concept regarding musical intervals is that, although we conveniently refer to musical intervals as a linear distance between two notes measured in *steps* or *half-steps*, the true underlying mathematical source of an interval's character is the *ratio* between the two frequencies. In other words, a perfect fifth sounds the way that it does because the top note vibrates one-and-a-half times faster than the bottom note. A note that vibrates exactly 2 times as fast as another note is considered to be an octave higher.

Here is a table showing the 12 chromatic intervals and the ratios they typically represent:

Chromatic Interval Ratios
(Just Temperament)

Interval Name	Symbol	RATIO	Decimal
Root (Unison)	R	1:1	= 1.0000
Minor Second	m2	25:24	= 1.0417
Major Second	M2	9:8	= 1.1250
Minor Third	m3	6:5	= 1.2000
Major Third	M3	5:4	= 1.2500
Perfect Fourth	P4	4:3	= 1.3333
Tritone	#4 or b5	45:32	= 1.4063
Perfect Fifth	P5	3:2	= 1.5000
Minor Sixth	m6	8:5	= 1.6000
Major Sixth	M6	5:3	= 1.6667
Minor Seventh	m7	9:5	= 1.8000
Major Seventh	M7	15:8	= 1.8750
Octave	8ve	2:1	= 2.0000

Ratios of smaller integers (not to be confused with smaller ratios of integers) are more "at rest" or "in tune" than ratios of larger intervals. For example, the major seventh (15:8) has a strong *need to resolve* upwards to its closest neighbour, the octave (2:1). The octave sounds more restful than the major seventh. A state of balance begs to prevail, and the major seventh becomes magnetically attracted to the octave - not unlike a satellite settling into orbit.

The sharp four and perfect fifth relate to each other in the same sort of way; the elastic tension of this specific interval combination can be heard in the first few notes of "Für Elise" by Beethoven. 45 and 32 are larger integers than 3 and 2, and are therefore the more dissonant musical pair.

This entire ratio-oriented approach, including its preference for smaller integers on both floors of the fraction, is one approach to the creation of a scale with "Just Temperament."*

The tendencies of these numbers to "boil down," and the manner in which they are creatively summoned, manipulated, satisfied, or avoided, are all context-dependent, and an integral factor** in the "personality" or "emotion" of any melody.

The tritone, with its abrasive high-integer ratio of 45:32, was once considered to be "evil" by a few freaky zealots with enough power to have it officially banned! Evil or not, it has become very widely used in many different styles of music,*** in ways that are generally considered pleasant and useful far beyond disposability or ethical questioning.

The poor old tritone has gotten a bad rap, if you ask me.

And, changing the topic slightly, so has the number 666. Have you ever played "Yahtzee"? Too many double sixes in a row are considered bad luck in that game. It's a fun game, but put yourself in six's shoes for a second...

After all, two-thirds is the fractional representation of the musical perfect fourth, which in decimal form is:

0.666666666666666666666666666666666666
666666666666666666666666666666666666
666666666666666666666666666666666666
666666666666666666666666666666666666
666666666666666666666666666666666666
666666666666666666666666666666666666
666666666666666666666666666666666666
666666666666666666666666666666666666
666666666666666666666666666666666666
666666666666666666666666666666666666
666666666666666666666666666666666666
6666666666666666666666666666...

The perfect fourth is the mirror inverse of the perfect fifth, which cannot be rightly considered "wrong" unless all of music and therefore the harmony of the entire universe is likewise considered evil. Any and all melodic and harmonic colourings which happen to exceed the

---

* Look up Pythagoras in the encyclopedia sometime! He & his students were totally gung-ho on all this integer-ratio musical interval stuff. You could almost say they invented it, but Mother Nature might get a little miffed about that sort of presumption.
** Pun almost intended.
*** The intro sections of "Purple Haze" by Jimi Hendrix, "Black Sabbath" by Black Sabbath, and "Pussywhipped" by Stormtroopers Of Death are all great examples of the tritone.

complexity of simple octave are unholy, is that it?

Ha! Utter cow-dung, as far as this authour is concerned.

I also found this nifty formula which will hopefully serve to further demystify and defend 666:

$$[\sin 666° + \cos (6 \times 6 \times 6)°] = \Phi$$

The Golden Number can't be wrong; therefore, 666 receives automatic mathematical immunity from any charges of terrorism, etc.

In any case, I just want to be perfectly clear that the mighty, incredible perfect fifth - represented by the fractions 3/2 and its inverse 2/3 as well as any doubling or halving of these (such as 3/4 or 8/3) - is commonly known as the *power chord*.

Pop culture seems to think that the guitar player from the Who (or maybe the Kinks or some other of our favourite bands) invented the power chord, but the vanity of humanity knows no bounds.

Nature invented it!

Power chords are extremely common in all sorts of music - not only rock and roll. Even the planets sing power chords... one Venus day equals two-thirds of an Earth year, and the orbital periods of Venus and Mars harmonize a power chord as well. Archimedes, a famous and ancient numbers person, was so impressed by the fact that the volume of a cylinder with its height equal to its diameter is 3/2 times the volume of the sphere that fits tightly inside, that he demanded the graphical depiction thereof be etched on his tombstone! Apparently, he also discovered the power chord ratio in the area of a parabola - but don't ask me to explain it 'cause it's a little out of my intellectual reach until I go get a few more math classes under my belt. I just had to mention it 'cause it's way too cool.

Thanks to rock and roll musicians throughout the ages for carrying the torch and keeping humanity in touch with one of the Universe's prime generators.

*Power chord pervades all.**

---

* Makin' me some pancakes for supper, I was commanded by the instructions to combine 3/4 of a cup of water with 1 full cup of pancake mix. That's the power chord ratio!

CHAPTER Φ

Perfect Fifth To The Power Of Twelve

So... to recap, we've seen two manners of representing the 12 chromatic intervals so far: 1) distances measured in semitones, and b) a collection of low-integer ratios.

Another method of deriving the exact frequencies for the Chromatic Scale is by reiterating the perfect fifth. Starting at a root note of C, for example, one would determine the next note by moving upward in pitch by the amount of a perfect fifth. Counting through the alphabet like so:

C=1, D=2, E=3, F=4, G=5

...then the fifth above C is G. In number-world, this means multiplying the frequency of the tone C by 3/2 (see the chart in Chapter Φ). Please observe the following calculation:

$$261.63 \text{ Hz} \times 1.5 = 392.45 \text{ Hz}$$
$$\text{C} \quad \text{up P5} = \quad \text{G}$$

Therefore the pitch of G is 392.45 cycles-per-second. Counting again alphanumerically: G=1, A=2, B=3, C=4, D=5; so the next fifth up from G is a D note.

$$392.45 \text{ Hz} \times 1.5 = 588.67 \text{ Hz}$$
$$\text{G} \quad \text{up P5} = \quad \text{D}$$

To make our collection of frequencies fall within the convenient range of one octave, we ought to divide D's frequency of oscillation (and any others which are larger than twice the root) by 2:

$$588.67 \text{ Hz} \; / \; 2 = 294.33 \text{ Hz}$$
$$\text{D} \quad \text{down 8ve} = \quad \text{D}$$

So... In three short steps, we've determined the frequencies of three notes by using the perfect fifth ratio of 3:2 as a multiplicator:

C = 261.63 Hz
D = 294.33 Hz
G = 392.45 Hz

Arranging these note letters clockwise around a circle in the same order that they unfold from the root via reiteration of the perfect fifth, a powerful construct is revealed. Its name is the "Circle of Fifths" and it's very popular and useful as a compositional tool, learning device, memory aid, mandala, and conversation piece:

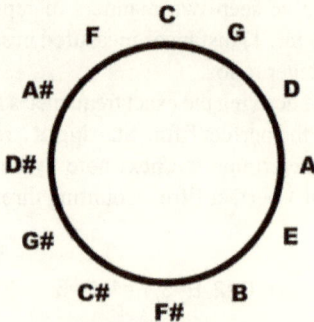

Circle Of Fifths

The fruits of this procedure can be more universally expressed in terms of fractions, since the C note's root frequency of 261.63 Hz is, in fact, arbitrary. Therefore, establishing a starting ratio of 1:1 will seed a collection of twelve fractions which can be subsequently multiplied by any desired root frequency...

Three-over-two times 3/2 equals 9/4, reduced by an octave to 9/8 to keep within a convenient one-octave limit. Continuing the process... 9/8 times 3/2 equals 27/16, or 1.6875; this is a bit out of tune from the major sixth 5/3 ratio of 1.66666, the closest match from the Just Temperament system. So, already we can see that this process will generate a set of ratios differing from Just Temperament. Continuing...

Unison	= 1 / 1	=	1.0
P5	= 3 / 2	=	1.5
M2	= 9 / 8	=	1.125
M6	= 27 / 16	=	1.6875
M3	= 81 / 64	=	1.265625
M7	= 243 / 128	=	1.8984375
#4	= 729 / 512	=	1.423828125
m2	= 2187 / 2048	=	1.06787109375
m6	= 6561 / 4096	=	1.601806640625
m3	= 19683 / 16384	=	1.20135498046875
m7	= 59049 / 32768	=	1.802032470703125
P4	= 177147 / 131072	=	1.35152435302734375
8va	= 531441 / 262144	=	2.027286529541015625

SECTION III: MIGHTY ALL-PERVADING POWER CHORD

...all 12 ratios of a Chromatic Scale are determined.*

Some interesting patterns appear. With each move forward a fifth, two more decimal places are added, and they all end in 25 or 75. The octave isn't even a perfect two-point-oh.**

Oh! An imperfect octave? Wierd.

This is indeed quite different from the Just Temperament system described in the previous chapter. This system of repeated fifths is referred to as the Pythagorean system of intonation, whether he actually invented it *first* or not. Ancient Chinese mathemamusicians also knew of this method. Is anything ever really *invented,* or only discovered and re-discovered? If time repeats itself, then *all of this* has happened before - several times.*** But I shan't get tangled in words. Let's just say he "discovered" it, about 2600 years ago.

Pythagoras was technically Greek, and was born in the sixth century B.C. He traveled around a lot and learned a lot. He hung out in Egypt for awhile - when it was still sort of ancient - so maybe he got his wisdom from aliens. His students were sworn to secrecy and not allowed to write anything down or eat beans.**** Even though he was always in search of harmony and truth, a few conservative wankers supposedly considered him dangerous, and he died at the ripe age of almost-a-hundred when his school was burned to a crisp.*****

But I could have sworn I saw him rummaging through garbage pails downtown the other day. Anyway... he also pioneered the *Pythagorean Theorem,* which is all about triangles and stuff. If memory serves, it goes something like, "The hippo tunes his triangle to the short side of the second square to the right of the longest side of the protractor." Or something like that. He is also credited as the originator of Western numerology.

I think a lot of these people were way ahead of their time or, like, maybe us modern people just don't learn enough of this stuff. I guess nobody cares. And why should we care about the structure of the cosmos, with all these videogames, beer, free porn and other various distractions at our disposal?

---

* Refer to Appendix 364.5 for a cross-discplinary correlation involving repeated fifths, of my own discovery.

** The discrepancy between the "true" octave and the "reiterated fifth" octave was discovered by Pythagoras and became known as the "Pythagorean Comma."

*** Have you ever noticed that both "learning something new" and "remembering" seem to share the same tone of a "light bulb going on above the head"?

**** Especially the ones shaped like testicles.

***** Rumours of suicide have also milled. According to Rashid from *Cowboy Bebop: The Movie,* it was poison beans that got him.

CHAPTER Φ

Temper, Temper!

As mentioned earlier, the Chromatic Scale divides the octave into 12 equal pieces. Oops, that's a bit of a lie... To be more precise, I should say 12 *approximately* equal pieces. Which is actually *less* precise. Ummm, what I mean is, equal is *less* precise...

I'll explain. As mentioned earlier, one ideal measure of a musical interval is contained in its *ratio,* and Just Temperament embodies this approach. Although Just Temperament is a very harmonious system, it's extremely high-maintenance from the performer's perspective; musical instruments would need to have their tuning adjusted a little differently every time there's a key change because of the various undulations which occur in the numbers that lie beneath the sound. This finicky aspect of natural harmony is so troublesome that humans bothered to invent things like a harpsichord with fifty-three keys per octave, in attempt to accomodate true harmony in multiple keys without the chore of re-tuning.

What a schmozzle!

Eventually, the "Equal Temperament" system was invented as a more versatile, but less harmonious, alternative to Just Temperament. As the name implies, the equally-tempered Chromatic Scale divides the octave into 12 *precisely equal* pieces. This allows a musical instrument to have a fixed tuning which works equally well for every key.

There is a serious trade-off, however: although Equal Temperament wins on the convenience factor, it is significantly compromised in the area of natural mathematical resonance. This wonky, out-of-tune system has become the standard throughout most of the world,* thanks in part to piano manufacturing setting the standard...

No wonder the sky is falling! Luckily, Just Temperament is still an option to some extent. For example, a violinist or lap-steel guitarist can adjust the pitch of any note, in realtime, with potentially high precision in such a way that every note will sound harmonious and natural. This is because these instruments are designed to allow for continuous variability in pitch. Singers and tromboners have the same luxury.

Unfortunately, guitarists have more serious hurdles on their hands. Because the Equal Temperament system is so badly out of balance, it

---

*...the *Western* world, at least.

can be extremely difficult to get the top two strings (B in particular) to sound adequately tuned for the open chords of both A major and C major, for example. Also, thirds and sixths sound strangely inconsistent. All of this is because the guitar's strings have a jog in the tuning pattern (fourth, fourth, fourth, *third*, fourth) and also because Equal Temperament is so out of whack. One way around this glitch in the guitar's symetry is to split the difference on the tuning of the B-string's comformity with the G-string versus the E-string, or avoid playing those two strings, or avoid playing thirds (power chords suffice for many styles anyways) or maybe just avoid the guitar altogether. What a nuisance! Do I really have to bend every note into the right pitch angle? I guess I could just make really out-of-tune sounding music on purpose, which I've been known to do... Spectralism to the rescue!

Okey dokey, well, this overly-political discussion with myself is starting to make me cranky, so I'd better stop. It's not like I can just make Equal Temperament go away, so I might as well stop complaining and just deal with it.

To quote a sax player named Gerry Mulligan:

> "The best part about playing the piano is that you don't have to lug around a saxophone."

Here's a table showing Equal Temperament's errors compared to Just Temperament:

Deviation Of Equal Temperament
(from Just Temperament)

Note	Just Hz	Equal Hz	Error (Hz)	Error (cents)
C	261.63	261.63	0	0
C#	272.54	277.18	+4.64	+29.23
D	294.33	293.66	-0.67	-3.94
D#	313.96	311.13	-2.84	-15.68
E	327.03	329.63	+2.60	+13.71
F	348.83	349.23	+0.40	+1.98
F#	367.92	369.99	+2.07	+9.71
G	392.44	392.00	-0.44	-1.94
G#	418.60	415.30	-3.30	-13.70
A	436.05	440.00	+3.94	+15.61
A#	470.93	466.16	-4.77	-17.62
B	490.55	493.88	+3.33	+11.72
C	523.25	523.25	0	0

Exhausting as it may be, this discussion of temperament and scale theory is by no means exhaustive. If you do the research you're likely to find tons of kwazy talk about schismas and edos, porcupine generators, syntonic commas, undecimal diesis, logflat badness, convex bodies, Fokker blocks, wedgies and what-have-you.

Happy hunting!

SECTION III: MIGHTY ALL-PERVADING POWER CHORD

CHAPTER Φ

Quincunx And The Legend Of The Chromatic Transformation Matrix Starflower

One fine catbird morning as I occupied the only cushy chair in the whole coffee-shop and contemplated the meaning of jazz and all that, I felt compelled to diagramize and arithmetize some sort of relationship between the Circle of Fifths to what I refer to with (purely platonic) affection as the Chromatic Circle.

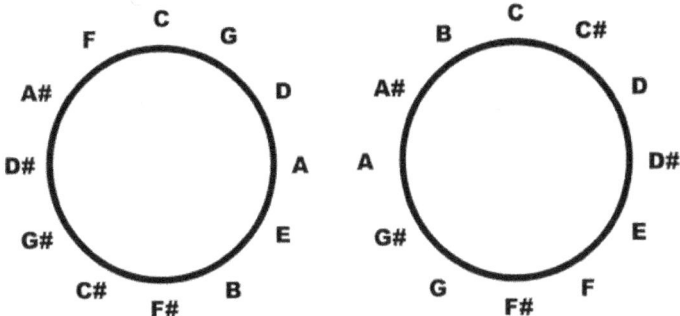

Circle Of Fifths   Chromatic Circle

The Circle of Fifths moves through the interval of a perfect fifth for every clockwise shift. The Chromatic Circle moves upwards a semitone for every clockwise shift.

How to "transform" one into the other...? That is the question.

But first, let's talk about *time*.

Time's progression is often shown as a horizontal line extending to infinity in both directions, with the past on the left and the future on the right. I hereby proclaim (with a presumed total lack of originality, I admit) that the circle does a better job of representing the true nature of time.

The Zodiac is quite circular. So is the clock. All things happen in cycles, so they say. The past is the future. History repeats its elves. An extension of the circle idea is the time *spiral* - a union of the timeline and the circle. Imagine corkscrew pasta. Events exhibits resonance not only through their angular position around the circle, but have the dimension of progression added to rotation.

Variation of themes - an evolution of revolutions.

In the spirit of the spiral, I spontaneously visualized a pattern in the Chromatic Circle while pondering the spiral nature of time:

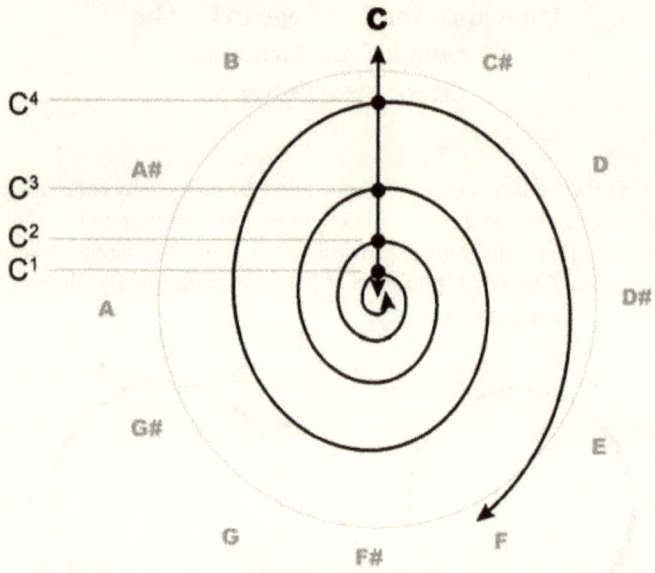

Chromatic Spiral

I'll explain the diagram above. The spiral represents chromatically modulating pitch. Imagine, if you will, a slide guitar, trombone, fretless bass, or a cello playing one long, sustaining note which gradually climbs from its lowest possible note up to the highest.* On the diagram this is represented by clockwise movement *outwards* around the spiral.

The vertical line, with "C" above it, represents a "C" note wherever it crosses the spiral. These points of intersection, marked by black dots, are labeled "C1," "C2" etcetera. C2 is an octave higher than C1, and C3 is yet another octave higher, etcetera. Spiraling outwards to higher and higher realms of vibration above and beyond the call of music for dogs, and inwards down to the lowest of bass frequencies comprehensible to only the most laid-back bassoonist.

I wonder if "etcetera" is Gibberish for "spiral"? Another question for the bank of etymological mysteries...

---

* Refer also to the classic ascending wail by Paul Di'Anno on "Wrathchild" by Iron Maiden! Robert Plant's rising glissando at the 7-minute mark of Zep's classic, "How Many More Times" is also a great illustration of a chromatic spiral.

SECTION III: MIGHTY ALL-PERVADING POWER CHORD 47

And, I also wonder about the treble clef.

Don't you?

It looks like it may be some kind of arcane encryption, a "spiral statement" of sorts.

Coincidence... or conspiracy?

Anyway, back to the comparison between the Circle of Fifths and the Chromatic Circle. Mathematically, the two approaches to systematizing frequency are quite different. The Circle of Fifths proceeds clockwise through multiplication by the ratio of 3-over-2, or 1.500 in decimal form. The equally-tempered Chromatic Circle, however, moves stepwise through multiplication by a factor of the 12th root of 2 (1.0594630944....).

I never thought of it until now, but a simple division of these two factors ought to yield the defining numerical relation between the two circles. In other words, the perfect fifth divided by the equally-tempered chromatic semitone is equal to:

$$1.5 \; / \; 2^{1/12} \; = \; 1.4158114690225402449628697350013...$$

This numerical specificity is something I just figured out just now. Although the discovery should feel like an achievement, it's really not too exciting compared to the spiral vision I came up with in the coffee shop that morning a few moons ago.* I was also perusing astrology at that point in my life (about 8 in the morning, I think), and was learning a little about the different *angular* relationships between the planets,

---

* Whoah - stop the presses! Brainfart alert: I just noticed that 1.4158114690225402449628697 is in dangerously close proximity to the square root of 2! Harsh. People have died over this stuff in the past. Many years ago, rumour has it, a Pythagorean disciple named Hippasus discovered that the square root of 2 is an irrational number (that which cannot be expressed as a ratio of two integers). Pythagoras, believing that everything in the Universe was expressible in terms of rational numbers, couldn't handle the ugly truth of root 2, and sentenced Hippasus to death! The square root of 2 is approximately 1.4142135623730950488016887. Yikes. I'm scared. Maybe I need to use a nom-de-plume for this book.

and the names and meanings of some of those angles.

In the Zodiac there are 12 "houses" (Ophiuchus notwithstanding)*
- Pisces, Aries, Taurus, Gemini, Cancer, Leo, Virgo, Libra, Scorpio,
Sagittarius, Capricorn, and Aquarius. In the Chromatic Scale there are
12 notes. Great! So, obviously there's some kind of connection to be
explored between music and astrology.

I'll discuss this in more detail a little later in this Section. Be
prepared to meet the *quincunx* and its homies.

Zodiac's got the curves, baby, I got the angles.

Angularly speaking, it takes 7/12 of a full rotation to equate the
Circle of Fifths to the first step of the Chromatic Circle. Another helpful
diagram to explain the phenomenon:

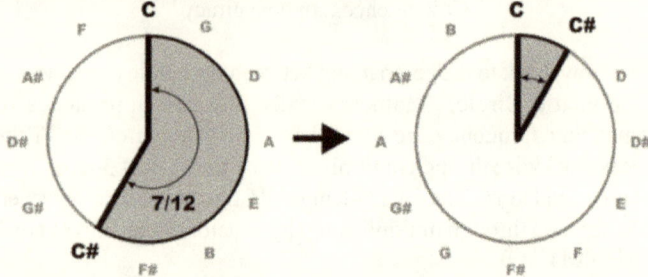

Circle Of Fifths   **Chromatic Circle**

Amazingly enough, transformation in the other direction uses the
same angle:

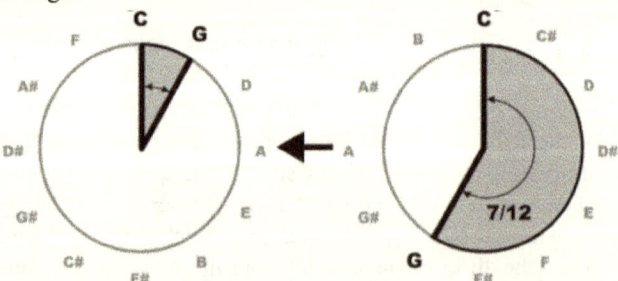

Circle Of Fifths   **Chromatic Circle**

---

* Apparently there's more than one way to look at the stars. Ophiuchus,
formerly known as Serpentarius, is one of the 88 constellations (hey - the
number of keys on a standard piano) and the only zodiacal constellations
not included as its own sign in astrology. Ophiuchus sits between Scorpio
and Sagittarius. See Appendix 13.

SECTION III: MIGHTY ALL-PERVADING POWER CHORD 49

If we repeat this angular movement, a pattern emerges...
Look... a star flower!

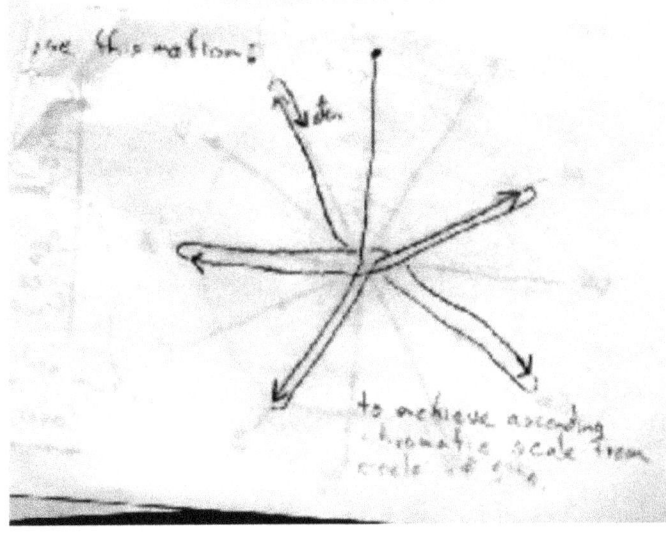

FIG. 4: A GENUINE SCAN OF ONE OF THE SKETCHES I MADE THAT MORNING AT THE COFFEE SHOP

The diagram above only goes through five ripititions of the cycle, which would ultimately require a total of 12 angular shifts before completing the round trip. On the next page is a cleaner-looking, step-by-step diagram of the *Chromatic Transformation Matrix Starflower* procedure...

The angle shown by each bent arrow is equal to 7-over-12 multiplied by 360 degrees, which works out to 210°. Or, an inside angle of 150 degrees (360 minus 210).* Seven of one, half dozen of the other.

The number 7 might itself be worth a little pondering in this situation, since there are 7 notes in your everyday average diatonic scale.

...not to mention 7 days in a week, 7 holes in the head, 7 deadly sins, 7 seas of Sinbad, 7th son of a 7th son, 7 and the ragged tagger, 7 priestesses with 7 trombones, 7 levels of hell, 7 lions of Ishtar's chariot, 7 chakras, 7-headed demons, 7 dimensions of existence, the turquoise lady's 7 faces, 7 pillars of wisdom, 7 ray energies of the Great Bear stars, Dance of the Seven Veils... parallels for future contemplation, perhaps. There must be some way to combine the jist of these structures with musical information; the way I see it, every number has a resonance and a collection of overtones, just like a musical note.

---
* Twelve minus 7 leaves 5 left over, which alludes to the *pentatonic* scales.

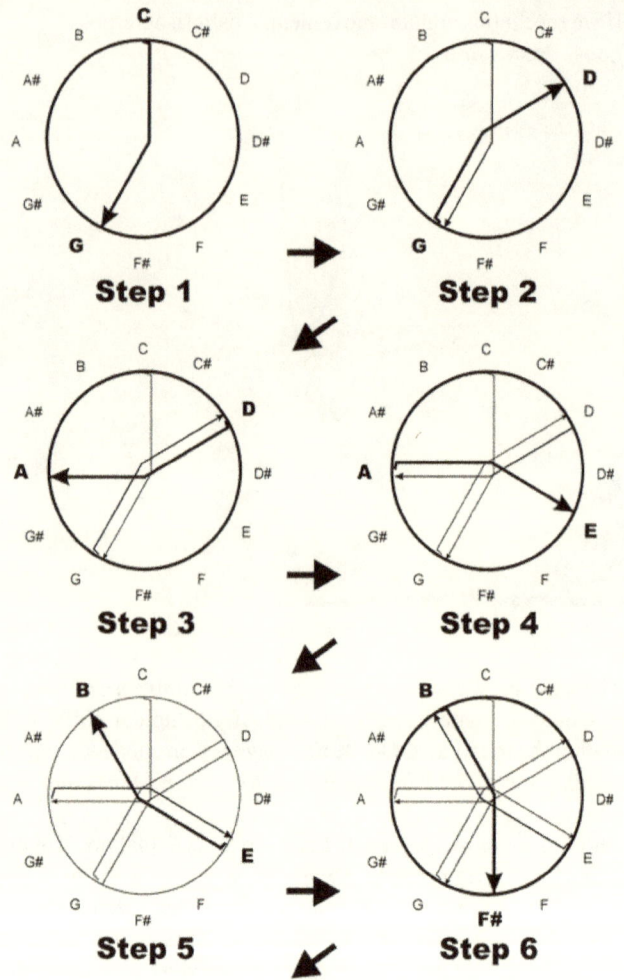

etcetera up to 12 steps and beyond

But, alas, I digress.

The Zodiac, which seems to have high potential for metaphoric or other levels of comparison with the Chromatic Scale because of its circular, cyclical nature and our division of both into segments of twelve and which hath thusly inspired the commencement of the writing of this entire chapter, prescribes specific names for the angles involved, just like music's naming of intervals.

Here are some of the names of these angles or "aspects" as they are called in astrology: conjunction (0°), sextile (60°), square (90°), trine (120°), opposition (180°), and quincunx (150°). These are called the

SECTION III: MIGHTY ALL-PERVADING POWER CHORD

"major aspects," and there are some "minor aspects" as well: semi-sextile (30°), semi-square (45°), sesquiquadrate (135°), quintile (72°) and bi-quintile (144°). The heavenly bodies move into, through, and out of these positions as they whirl around the Sun like an orchestra of dervishes.

The angular expression of the Chromatic Transformation Matrix Starflower is one-hundred-and-fifty degrees, which therefore equates it to the zodiacal quincunx.

Another observation: the quincunx ratio of seven divided by twelve is equal to 0.58333333333333333333333333333333333... That's fairly close to phi, which is 0.6180339887...

The other closest candidate would be eight. Eight divided by twelve equals 0.666666666. Which is closer?

Seven-over-twelve compared to phi:

```
0.5833333333  /  0.6180339887   =  0.9438531601
    100 x (1  -  0.9438531601)  =  5.6% error
```

Eight-over-twelve compared to phi:

```
0.6666666666  /  0.6180339887   =  1.0786893258
    100 x (1.0786893258  -  1)  =  7.9% error
```

Seven wins!

There you have it - of all the possible divisions of twelve, seven is closest to the Golden Mean. By this method of measurement at least, the astrological quincunx is the *golden aspect*.

All hail the glorious quincunx, for it shall transform us from chromaticism into the mighty perfect fifth and back again, on and on into the great infinite.

I'm sure there are other ways to equate the twelve pieces of Zodiac pie with the Chromatic Scale and other areas of music theory. The quest to find them is yours, should you choose to accept it.

Stay tuned for more Zodiac music theory at the end of Section III.

CHAPTER Φ

The Palindromic Di-Componential Chromatic Scissional Rotation Procedure

The following procedure is something I discovered through ponderance and experimentation. If we separate the Chromatic Circle into two components by alternating notes, we get the following arrangement:

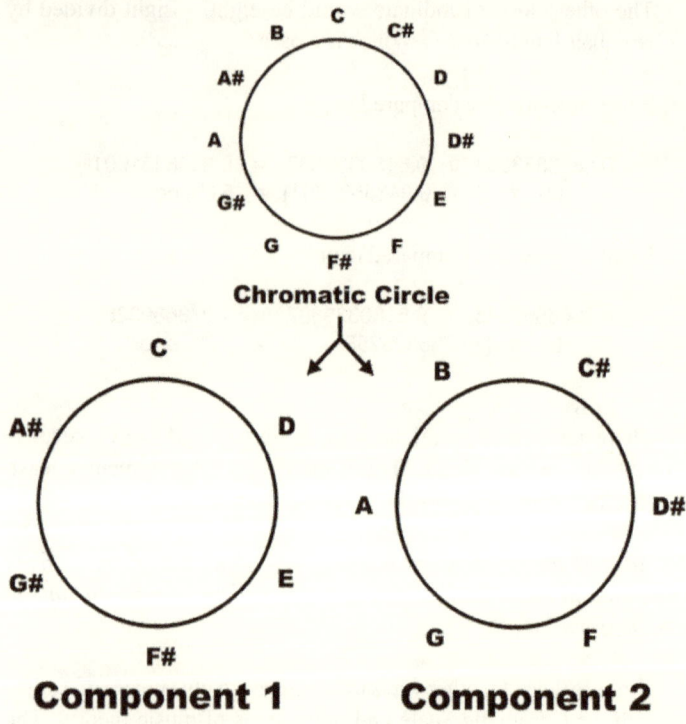

Each component is a series of whole steps - a.k.a. the *whole-tone scale*.

Next...

Notice how, by rotating component 2 by 180 degrees and re-joining it with component 1, the Circle of Fifths is generated.

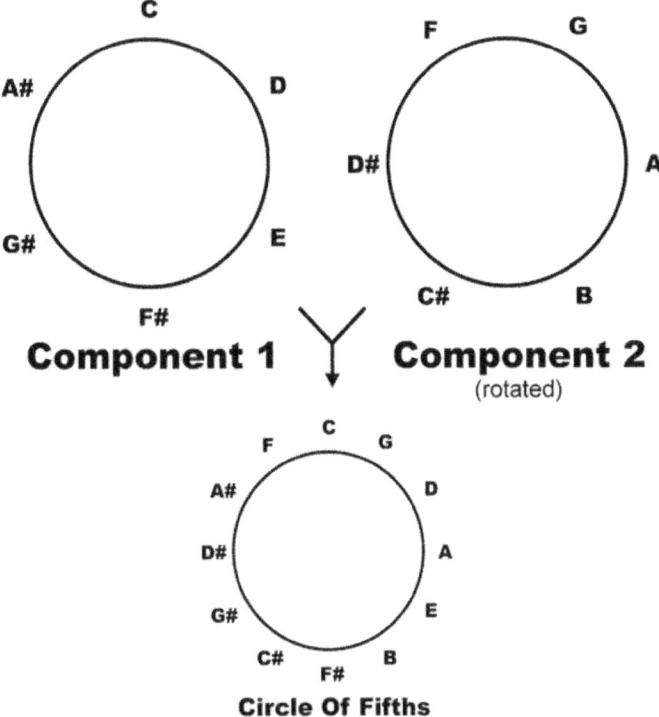

Circle Of Fifths

This procedure is palindromic; that is to say, it works the same in the opposite direction - the Circle of Fifths transforms back into a Chromatic Circle via the exact same procedure.

Neato, mosquito! Another observation: with all of the musical circles I've mentioned here, an angle of 180 degrees is equivalent to the diminished fifth, also called the tritone. I find this a bit of a surprise considering that a half-circle is so symmetrically simple and obvious in a geometric context, whereas the diminished fifth is the most and abrasive of all musical intervals.

Then again, considering that 12 semitones make up the Chromatic Circle and the tritone equals exactly half as many, it's no surprise at all. In fact maybe it explains why leading-tone modulations (which rely on the tritone contained in the dominant seventh chord) are so powerful and popular.

I haven't yet explored the zodiacal ramifications of the Palindromic Di-Componential Chromatic Scissional Rotation Procedure, but maybe someday I will. Or, *someone* will.

CHAPTER Φ

The Golden Interval

It occurred to me that the minor sixth ratio of 8:5 specified by the Just Temperament system is very close to the Golden Ratio. Eight over five equals 1.6000, only 1.11% below the Phi value of 1.6180339887...

By comparison, the just *major* sixth is 5:3, or 1.666666666666... This is on the high side of Phi by 3%, a less perfect match than the minor sixth.

Maybe the "true, optimum, perfect, natural" minor sixth is really defined by the ratio of Φ to 1, rather than 8 to 5. I wonder if this is already an established part of tuning theory and knowledge. If my proposition holds as much water as I think it does, the minor sixth is the *golden interval*.

[Remind me to tell you the story of the two-tone tuning fork some time...]

By the way: Equal Temperament's minor sixth is lower than Phi by 1.89%, somewhat less accurate than the just minor sixth. The just incarnation of the major sixth is also more accurate to Phi than the equally-tempered version.

This golden interval is a really great thing, and not only on paper. Lots of memorable songs use it. It often appears as the chord change E minor to C major. Or A minor to F major. My current theoretic rendition for the sake of this page is E major to E minor to C major to C minor to G-sharp major to G-sharp minor and back to E major - and through and through the cycle over and over again if that's your kick.

Enjoy.

CHAPTER Φ

Golden Arpeggios

Given that the minor sixth is closest to the Golden Mean out of the twelve choices from the chromatic intervals, it follows that the augmented chord is "the golden chord," and, spread out, it becomes a "golden arpeggio." Starting from E, a minor sixth re-iterated several times would create the note series E C G# E C G# E C G# E C G# E C G#... on and on in both directions, up and down. This is an inversion (ie. re-arrangement) of the spelling E G# C, which is a set of stacked major thirds. In music, an inversion of a chord is still essentially the same chord because the re-arrangements are all octave-based. Therefore, a series of stacked minor sixths is significantly equivalent to a stack of major thirds.*

A Tempered Golden Arpeggio

In the above diagram, each star gets smaller by a factor of exactly phi each time, to illustrate the idea of a perfectly symmetrical golden arpeggio. The stars get smaller as the notes go higher. I chose this representation because higher notes have smaller wavelengths.

I should point out an anomaly here: if the musical pitches were to match the quality possessed by the stars in the diagram, that of scaling by "exactly phi each time," then the names E, C, and G# would not hold water for more than a couple octaves or so. The ratio of 1.6180339887 to the power of 3 comes out to 4.2360679771, far less than perfectly aligned with the pure double octave of 4.0000000000. The 2nd E would sound rather F-ish and the other notes would also be too high. This misalignment would accumulate over subsequent evolutions of the

---

* Coming soon: "The Modal Mirror."

interval, and a consistent set of note names could no longer apply. Methinks that this sort of arpeggio, although it doesn't conform to the conventional scale-width of an octave, is actually a more correct and "pure" incarnation of the Golden Mean. Who knows - maybe the lifting of each octave has an uplifting effect on the listener...

CHAPTER Φ

The Times-Table Mirror

Penny universities...

Saturated in awe of the intricate patterning so prevalent in the world around me and faced with the luxury of exploring these realms, I found myself once again situated amongst the comfort of random human noise at another downtown coffee-shop, sipping tea with my calculator and notepad for hours on end.

Entwined in the fibres of ponderance and discovery, my mission on this occasion was to hunt down any musical treasures to be found in the times-tables.

First, I wrote out all 9 of the times-tables with the answers reduced to a single digit. The following example shows the 4x table:

```
4 x 1 = 4        SR = 4
4 x 2 = 8        SR = 8
4 x 3 = 12       SR = 1 + 2 = 3
4 x 4 = 16       SR = 1 + 6 = 7
4 x 5 = 20       SR = 2 + 0 = 2
4 x 6 = 24       SR = 2 + 4 = 6
4 x 7 = 28       SR = 2 + 8 = 10, 1 + 0 = 1
4 x 8 = 32       SR = 3 + 2 = 5
4 x 9 = 36       SR = 3 + 6 = 9
4 x 10 = 40      SR = 4 + 0 = 4
...ad infinitum
```

A repeating pattern of 4 8 3 7 2 6 1 5 9 was revealed by the 4x table. The rest of the times-tables gave the following repeating patterns:

```
1x:   1 2 3 4 5 6 7 8 9' 1 2 3 4 5 6 7 8 9' ...
2x:   2 4 6 8 1 3 5 7 9' 2 4 6 8 1 3 5 7 9' ...
3x:   3 6 9' 3 6 9' 3 6 9' 3 6 9' 3 6 9' 3 6 9' ...
4x:   4 8 3 7 2 6 1 5 9' 4 8 3 7 2 6 1 5 9' ...
5x:   5 1 6 2 7 3 8 4 9' 5 1 6 2 7 3 8 4 9' ...
6x:   6 3 9' 6 3 9' 6 3 9' 6 3 9' 6 3 9' 6 3 9' ...
7x:   7 5 3 1 8 6 4 2 9' 7 5 3 1 8 6 4 2 9' ...
8x:   8 7 6 5 4 3 2 1 9' 8 7 6 5 4 3 2 1 9' ...
9x:   9' 9' 9' 9' 9' 9' 9' 9' 9' 9' 9' 9' 9' 9' 9' 9' 9' 9' ...
```

In a zippy flash of eureka-gasm brainfartness I thought, "I will arrange each of these patterns into its own circular graph. I shall connect the dots!"

And so I connected the dots. I had imagined that some sort of nail-art type of imagery would appear, which it did. But some unexpected twists and turns emerged as well.

Firstly, I arranged the numbers 1 thru 9 clockwise around the outside of a circle, establishing a template for the graphs:

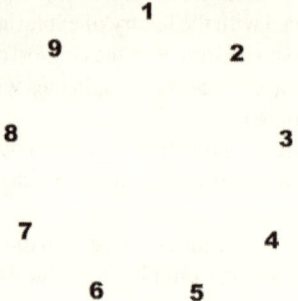

Next, I connected the dots around the circle in the skein order for each times-table. Using the 4x table as an example again (with the sequence 4 8 3 7 2 6 1 5 9 as shown on the previous page):

Cool looking pattern, eh?

But the real big surprise is the symetry in the *sequence of shapes* that are created through this procedure. Not including the 9x table, which seems to be reserved for its own special function as a sort of "mirror" or "grounding anchor" for the rest of the system, the series of 8 shapes *evolves palindromically*. They appear in the same order, backwards or forwards.

Also, polarity in regards to clockwise or counter-clockwise is reversed across the mirror. The 4x table shown above, for example, spins clockwise whereas its twin, the 5x table, is *counter*-clockwise.

SECTION III: MIGHTY ALL-PERVADING POWER CHORD 59

The following diagram shows the pattern of patterns:

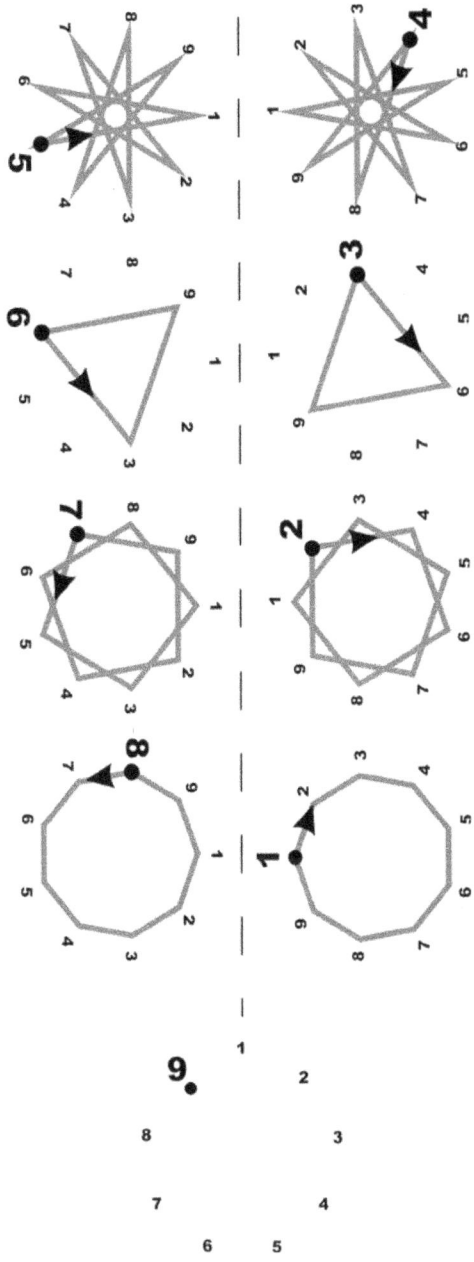

I originally had the diagram laid out lengthwise with all the shapes side-by-side, but I wanted it to fit on a single page without being too small to read. The folded arrangement has the added bonus of resembling a V-8 engine! The empty 9x graph at the bottom of the previous diagram resembles the engine's *distributor*, with the dot at 9 perhaps corresponding to the electrical supply wire. I wonder - what sort of pattern would emerge if the shapes were re-arranged to conform to the firing order of the sparkplugs of a real V-8?

Another idea for the deferral list, I suppose. Second edition... "Chaos In Boxes volume 2," perhaps?

[There's this little yellow spider that keeps crawling on the computer screen; when the typing cursor moves towards the spider, it jumps out of the way every time, to avoid being attacked by the cursor. Silly little spider.]

The pattern gains new beauty and grace when arranged as a circle of circles:

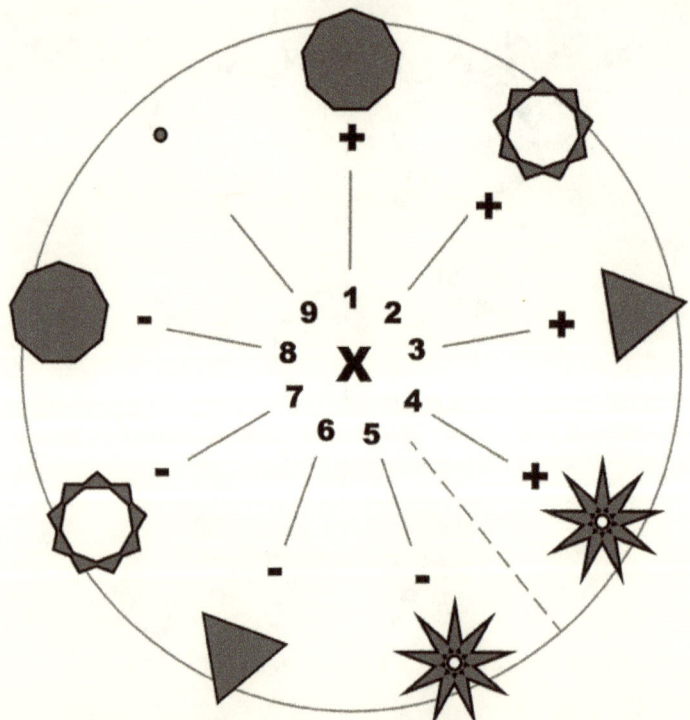

The Times-Table Mirror

Eureka!

SECTION III: MIGHTY ALL-PERVADING POWER CHORD

Eventually I noticed that, by folding the circle in half along the dotted line, the numbers match up a certain specific way: the nonagons are numbered 1 and 8; the flowers are 2 and 7; the triangles are 3 and 6; the spikeys are 4 and 5. Each number pair totals 9, which is also the mirror number of the entire system - what an incredible coincidence!

Even more amazingly and perhaps enough justification alone for the mere existence of this entire book, this "folding 9-sum" property is perfectly echoed in the bi-directional measurement of musical intervals and their sum of 9: according to standard Western music theory, a movement a *third upwards* is equivalent to moving a *sixth downwards* - and, of course, three plus six equals *nine*.

Also noteworthy is the flip-flop pairing of the major/minor quality of these interval additions. That is, an upwards *minor* third (eg. E to G) equals a downward *major* sixth (eg. E to G), whereas an upwards *major* third (eg. E to G#) equals a downward *minor* sixth (eg. E to G#). Comparison of this musical fact to the polarities in clockwise-ness found in each pair of times-table shapes confirms yet another new point of analogy perhaps significant enough for contemplation.

Number nine.
Number nine.
Number nine.
Number nine.
Number nine.
Number nine.
Number nine.
Number nine.
Number nine.

CHAPTER Φ

The Modal Mirror (Bumblebees & Basketballs)

Let's flip some scales *upside-down!*
Recall from the beginning of this Section that a half-step is one fret's distance on the guitar, and a whole-step is two frets (W = whole-step, H = half-step). Starting from its root note and moving upwards one note at a time, that old cliché the Major Scale has a stepwise spelling of W W H W W W H.

Its interval spelling: root, major second, major third, perfect fourth, perfect fifth, major sixth, major seventh. This can be abbreviated R, 2, 3, 4, 5, 6, 7; other scales would likely have modifications to this, indicated as "flat 2" or "minor 6" or "m7" or "sharp 4" or "#5" or other alterations.

With no sharps or flats, the C Major Scale is the simplest example to discuss in terms of note names: C plus a whole-step equals D, and D plus a whole-step is E, and E plus a half-step is F, and F plus a whole-step is G, and G plus a whole-step is A, and A plus a whole step is B, and B plus a half-step returns to C, completing the circle.

Well... how about starting at C and *minusing* the same pattern of steps, spelling *downwards* instead of upwards? C *minus* a whole-step equals B-flat. There is no B-flat in the key of C major... already apparent, a new scale is going to be generated that is different from the C Major Scale... a mystery unfolding!

Continuing the process:

$$C - W = B\text{-flat}$$
$$B\text{-flat} - W = A\text{-flat}$$
$$A\text{-flat} - H = G$$
$$G - W = F$$
$$F - W = E\text{-flat}$$
$$E\text{-flat} - W = D\text{-flat}$$
$$D\text{-flat} - H = C$$

This new collection of notes is commonly known as the Phrygian mode, and has a rather flamenco-ish sort of sound.

"Mode? What the heck is that? Fridge... Ice cream? Flamenco ripple? Huh?"

Modes are simply phase-evolutions of an existing "parent" scale, much like phases of the moon. A parent scale with a certain number of notes has the same number of "child" modes, each of which is also a musical scale. Every note gets a chance to be the root of its own mode.

Each of these modes exudes its own unique "mood" which differs from its parent scale as well as its sibling modes. The child and parent modes basically have an equivalent level of expressive sovereignty, to the extent that any of the modes could rightly be considered the "original from which the others are derived". However, one of them is usually designated as the originator.

"Mode 1" refers to the perceived originator. Therefore, mode 1 of the Major Scale (W W H W W W H) is simply *Do-Re-Mi-Fa-Sol-La-Ti-Do* itself. By definition, mode 2 starts on note 2 of mode 1 and has a shifted version of mode 1's step-pattern: W H W W W H W. Mode 3 begins on the third note of mode 1: H W W W H W W. This derivation process continues through all 7 possibilities:

```
mode 1:*          W W H W W W H
mode 2:**         W H W W W H W
mode 3:***        H W W W H W W
mode 4:****       W W W H W W H
mode 5:*****      W W H W W H W
mode 6:******     W H W W H W W
mode 7:*******    H W W H W W W
```

Each mode sounds different because its step pattern is unique, and therefore so is its collection of intervals. Through music, numbers exhibit resonance, colour and emotion. Mode vocabulary is extremely useful for composers and improvisers alike!

Although any conceivable scale is automatically a mode generator, when a musician talks of "modes" these days they're usually referring to the 7 modes of the Major Scale, listed in this page's footnotes.

So...

That's enough background theory for now; back to the process of flipping scales upside down! When Ionian is inverted, Phrygian mode is the result. An intriguing transformation indeed!

I went ahead and flipped each of the 7 modes to establish their "undertone" counterparts. The results amazed me, especially when plotted on a circular graph. With the mode numbers evenly spaced sequentially around a circle, the flipperoos make a bunch of parallel lines going back and forth across the circle.

---

* a.k.a. Ionian: R, 2, 3, 4, 5, 6, 7.
** a.k.a. Dorian: R, 2, m3, 4, 5, 6, m7.
*** a.k.a. Phrygian: R, m2, m3, 4, 5, m6, m7.
**** a.k.a. Lydian: R, 2, 3, sharp 4, 5, 6, 7.
***** a.k.a. Mixolydian: R, 2, 3, 4, 5, 6, m7.
****** a.k.a. Aeolian: R, 2, m3, 4, 5, m6, m7.
******* a.k.a. Locrian: R, m2, m3, 4, flat 5, m6, m7.

The arrangement resembles a bumblebee, or maybe a basketball:

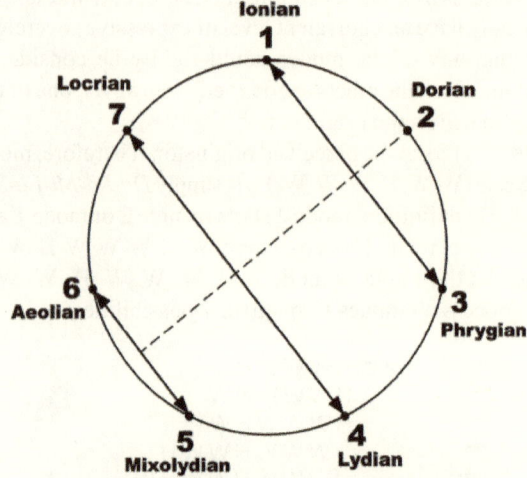

The Modal Mirror

Curiously, the Dorian* mode remains the same when it's flipped upside-down! It also presides at the exact center of the system, at one end of the perpendicular to the parallels;** mode 2*** is the mirror axis between undertone and overtone, the glassy surface of the 7 tonal seas! With some practical experimentation, the expressive potential of jumping across the Modal Mirror becomes self-evident.

And so I pondered... what sort of results would come from inverting some other scale types? What about inverting the Blues scale, Harmonic Minor, Jazz Minor, Arabian, Lydian Dominant, Hungarian Gypsy, Enigmatic, Balinese, Algerian, Hirajoshi, Pelog, Hindustan, Ritusen, Piongio, Prometheus, Scriabin, Iwato, Egyptian,

---

* The modes were named after tribes of people in and around ancient Greece; the complete history of modes, and their technical specifications, is tangled and inconsistent, so I have only presented the "modern modes" here, and avoided the old "church modes" and the original (older) Greek modes altogether.
** The other end of the Dorian mirror is flanked by Aeolian and Mixolydian - each differing from Dorian by only one interval, further illustrating Dorian's "in-between-ness" with beautiful consistency.
*** Notice how, in Chapter Φ, "The Times-Tables Mirror," the number 9 is the axis of the system. In the Modal Mirrors, 2 is the axis. It just so happens that degree 2 of a typical 7-note scale is also referred to as 9, since 8 is the octave which is a duplicate of the root: 8=1, 9=2, 10=3, 11=4, 12=5, 13=6. *A direct connection between the times-tables and the 7 modes...* who knew? See Appendix 1497, "Palindromic Skein Sequence In The Squares" for more intriguing 9-axis stuff. ($0^2 = 9$?)

SECTION III: MIGHTY ALL-PERVADING POWER CHORD

Romanian Minor, Kumoi, Jewish, Mongolian, Locrian Major, and any others from the endless palette of scales?

By digging into the undertone side of things, the possibilities are suddenly *twice as endless* as they already were! [$2 \times \infty = \infty$.]

Exploring all of these inverted scales on the guitar could provide years of entertainment and study, if not lifetimes! Are we there yet? A short poem to celebrate the notion:

> Alas
> to be a vampire
> and   never   age or
> die
> !

Gotta start somewhere - how about taking a step towards simplicity by using only 5 notes instead of 7... What happens when the classic Minor Pentatonic scale (W+H W W W+H W) is inverted?

A Minor Pentatonic	= A C D E G A
upside-down A Minor Pentatonic	= A F# E D B A

A musical composition could feasibly concentrate on the upside-down Minor Pentatonic tonality for one section and then flip to the traditional upward spelling for a melodic contrast. Or, both at the same time! The above-water spelling could be manifested by a flute while the bass carries the inverted contour, for example.

Q: "What happens when *both sides* of the Minor Pentatonic mirror are combined into *one scale?*"

A: Collecting all notes from both of these 5-note scales, a new scale with 7 notes (A B C D E F# G) is produced - a more relevant answer than I was expecting, as these are the same notes as A Dorian - the exact same mode that functions as mirror-axis of the amazing Modal Mirror circle!

> Minor Pentatonic scale
> +
> inverted Minor Pentatonic scale
> =
> Dorian, axis mode of the Modal Mirror

Go figure!*

---

* I'd already been using the modes on the guitar for over 15 years before discovering these amazing symmetries - right under my nose all this time!

CHAPTER Φ

Rock Stars

There is a particular mirroring function contained in traditional astrology which enabled me to correlate the signs of the Zodiac with the notes of the Chromatic Scale - one of my longtime ambitions come to fruition!

Each Zodiac sign is ruled by a planet. Notice the mirrored arrangement:*

♈ = Aries
♉ = Taurus
♊ = Gemini
♋ = Cancer
♌ = Leo
♍ = Virgo
♎ = Libra
♏ = Scorpio
♐ = Sagittarius
♑ = Capricorn
♒ = Aquarius
♓ = Pisces

☽ = Moon
☉ = Sun
☿ = Mercury
♀ = Venus
♂ = Mars
♃ = Jupiter
♄ = Saturn

Notice that the ruling planets for Cancer and Leo, the two signs on either side of the reflection line, are the Sun and the *Moon*. The Moon is visible in the sky because it *reflects* light from the Sun. How appropriate... of course, my task was to superimpose the Modal Mirror system onto the Zodiac. It couldn't be simpler. Poetry in motion, these heavenly bodies are.

First, represent each musical mode by its parent tone. Assuming the key of C major: mode 1 = C, mode 2 = D, mode 3 = E, mode 4 = F, mode 5 = G, mode 6 = A, and mode 7 = B. Next, simply align the Dorian mirror with the ruling planet mirror, with a slight adjustment to overcome

---

* The planetary rulership system used by astrologers has changed over the years to include the outer planets, which were previously unknown. I'm using the older, simpler system because it illustrates the mirror more obviously.

the awkward inconsistency of "in-between-ness" (since the Zodiac has an even number of elements whereas there are an odd number of modes). Next, fill in the rest of the modes circle to create the Chromatic Circle; the 12 notes correspond perfectly with the 12 signs of the Zodiac!

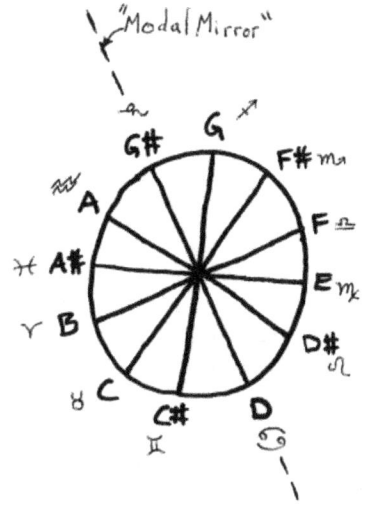

So:

C	(Ionian)	= Taurus
C#		= Gemini
D	(Dorian)	= Cancer
D#		= Leo
E	(Phrygian)	= Virgo
F	(Lydian)	= Libra
F#		= Scorpio
G	(Mixolydian)	= Sagittarius
G#		= Capricorn
A	(Aeolian)	= Aquarius
A#		= Pisces
B	(Locrian)	= Aries

The fretboard diagram to the right (a special bonus for the guitar players in this readership) illustrates the C Major Scale (position VII) with Zodiac symbols:

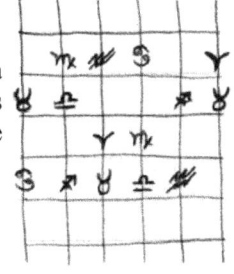

A variation of this system is to align Dorian with Leo instead of with Cancer, thereby favouring the Sun as the "centre of it all" instead of the Moon. In this case, the system shifts by one Zodiac sign:

C	(Ionian)	= Gemini
C#		= Cancer
D	(Dorian)	= Leo
D#		= Virgo
E	(Phrygian)	= Libra
F	(Lydian)	= Scorpio
F#		= Sagittarius
G	(Mixolydian)	= Capricorn
G#		= Aquarius
A	(Aeolian)	= Pisces
A#		= Aries
B	(Locrian)	= Taurus

Seven of one, half dozen of the other? I currently prefer the Cancer-centered system, because the Moon's reflective quality seems to bind the metaphorical tie more literally.

By accepting *both* versions of the alignment assignment, every Zodiac sign gets a chance to be a mode.*

All of these discussions have been assuming that "C" is the root. If another key was chosen - F# major, for example - the system would need to be adjusted so that F# = Ionian = Taurus, in order to maintain the alignment of the Modal Mirror and the ruling planet mirror.

Some might argue that a more obvious and therefore appropriate correlation is to align mode 1 with Aries, since Aries is the first sign of the Zodiac. In my mind, this method does not hold as much water as the mirror-alignment method. Therefore, in the interest of wide-spread consistency, I do proclaim and request that Western astrology be modified so that Taurus is counted as the first house, instead of Aries.**

A.S.A.P.
Stat.
S.V.P.
Thank you.

---

* Maybe Ophiucus could clear up the confusion here. Perhaps Ophiucus' position between Scorpio and Sagittarius should be aligned with Dorian somehow...
** And so I ponder... was the Age of Taurus an era more musically synchronistic than the present time?

Section IV

DEEP LIGHT

*Sound is vibration.
If that vibration starts to repeat itself,
a wave may be perceived and a musical tone is
the interpretive result.
Light also has a wave-like property.
Ripeness for comparison is imminent and/or eminent.
Are sound and light really one and the same?*

CHAPTER Φ

Light Equals Sound

The highness or lowness of a sound is commonly measured in terms of *frequency,* or "how many cycles occur over a period of time". Colours of light, on the other hand, are typically measured in terms of *wavelength,* or "how long it takes for one cycle to happen".

The two concepts are the inverse of each other - not unlike Yin and Yang or the the Circle of Fifths being compared to the Circle of Fourths.* In essence, wavelength and frequency are actually two different ways of talking about the same thing: a high frequency implies a short wavelength, just as a low frequency implies a long wavelength.

The chicken is the egg. Flashlights shoot sonic booms. Dog-whistles pour rainbows. The fat lady sings and the lights dim. The crowd goes wild.

By only a slight stretch of the imagination, music is a beam of light spoken in a very deep voice. All divisions, expressions, mediums and singularities are connected - all the same one thing. *Hmmmmmm...* So simple, yet profound.

Even though hearing and seeing are two different things for us humans, the realms of light and sound could be considered as one continuous spectrum. Scientific investigation has institutionalized various names to describe the various wavelength regions, both above and below the realm of light - all the way down to *sound... and beyond!*

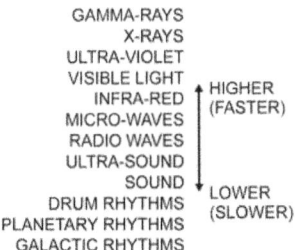

By adequately cranking up the speed of a comparatively slow waveform, such as a drum rhythm, a constant drone could be created. Spin the record even faster and light would eventually come pouring out from the speakers. I insist.

Ears and eyes are separate, while sound and light are one. Perhaps

---

* See Appendix 4, "The Circle Of Fourths." Are you looking in the mirror, or is the mirror looking at you?

the separation of the one spectrum into two separate transducers was one of God's slapdash, Mickey-Mouse solutions to a biological engineering pickle that even (s)he couldn't handle. Maybe some distant alien peers feature an integrated biological design which uses a single sensor to receive sound, light, smell, taste, touch, thoughts, and any other imagineable array of sensations...

A poem to celebrate the equation:

<pre>
       light
        is
       sound
        is
        god
        is
        dog
        is
        fly
        is
       frog
        is
       whale
        is
       guppy
        is
       puppy
        is
       kitty
        is
     itty-bitty
        is
      infinity
</pre>

The glaringly unanswered question in my personal repertoire of useable knowledge: what colour is each musical note on my guitar? Conversely,* what note is sung by each colour of the rainbow? Upon these ponderances, I resolved to quantify these relations between colour wavelengths and sound frequencies.

One could apply this information in a variety of ways. A song in the key of C-sharp might have a video with lots of green, for example. Other applications of the synergistic synchronization between colour and sound could include various sorts of meditation, therapy, memory enhancement, past-life hypnosis, lucid dream induction, building demolition, earthquake prevention, weather manipulation, etcetera.

The Circle of Fifths and the Chromatic Circle would be more

---

* Same question, only upside-down.

SECTION IV: DEEP LIGHT

pleasant to look at if they were coloured and, if a relevant standard could be established, they would be much more enlightening as music teaching tools. The collective musical experience could become more integrated.

This is what I came up with, using a somewhat scientific method:

Objective:
To derive musical frequencies from the wavelengths of the various colour ranges of light.

Method:
1) Get on the fast track by using the Internet to find the wavelengths for R, O, Y, G, B, and V. I'm not too sure what to say about the missing "I". None of the sources I found seem too include Indigo. Is this some sort of conspiracy? Poor old Indigo has gotten a bad rap, if you ask me.
2) Convert all wavelength data to frequency, which is a more music-friendly term than wavelength.
3) Transpose the frequency down by several octave increments to preserve the musical "tone" of each colour. Contrive a simple formula which includes the necessary number of octaves of downward transposition.
4) Graph the findings for possible future application.

Observations:
After finding several, slightly differing sets of data regarding wavelengths of colours, I settled on this arrangement (expressed in nanometers):

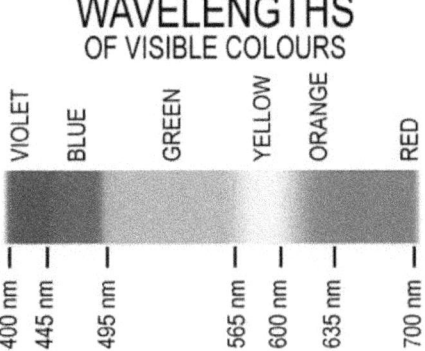

Nano, nano. Shazbot!
Using the secret magic formula, convert wavelength to frequency (measured in Hertz,* a.k.a. cycles-per-second) and transpose into audible range:**

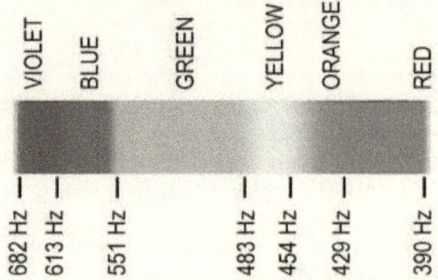

Next, express the frequencies in musical terms:

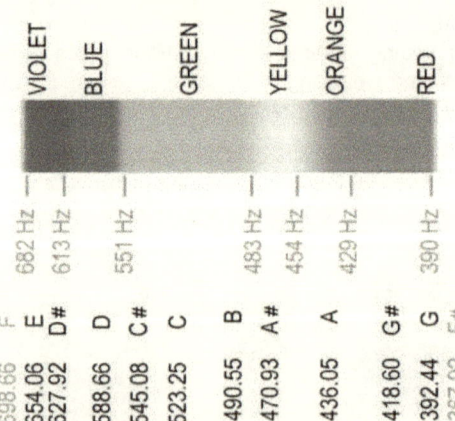

Hey! What the hell is going on around here? There are notes missing! F and F-sharp don't even fit

---

* See Appendix 5, "Hertz Donut."
** See Appendix 272659.6521 for secret formula.

inside the spectrum! I guess they're in the non-visible ranges of ultraviolet and infra-red. Even G is right on the verge of invisibility. Apparently, our range of visibility is less than an octave's worth of spectrum - a strangely disappointing revelation. What a ripoff - not even a full octave of visual bandwidth!?*

Have you ever heard of the *interrobang?* It's a hybrid combination between a question mark and an exclamation point. It doesn't show up as standard equipment on these here computer keyboards, but... someday, maybe. The interrobang appeared briefly on a few typewriters (way back in the days of typewriters), and it *is* available in some fonts as a special character.

But, alas, I digress...

If we were spiders maybe we could see F-sharp notes. Some animals can see ultraviolet and/or infra-red, if I'm not mistaken. But us humans are stuck with less than an octave. Oh well. I guess we can't be the superior species in *every* aspect! (Joke.)

Our range of hearing, on the other hand, is much more than a mere octave. The human range of hearing is about 20 Hz to 20,000 Hz at best (I tested my own range using a wave editor and it was not quite so wide - up to about sixteen grand or so, and probably diminishing with age). A special math whatchamacallit known as the *logarithm* may be employed to derive the equivalent range expressed in octaves. According to my calculation, the human range of hearing is almost 10 octaves:

$$20,000 = 20 \times 2^x$$
$$1,000 = 2^x$$
$$x = \log_2 1000$$
$$x = 9.9657842$$

One redeeming quantitative dimension for our visual faculty is that of sensitivity. Apparently, it only takes about 5 or 6 photons in a

---

* One day walking home from the rainy park, two rainbows hovered in the sky. Not only were they concentric, but they were nestled so close together that they were touching, like a single rainbow with a repeated spectrum (R O Y G B V R O Y G B V). This seemed to me like the reiteration of a musical scale over a 2-octave range: a 2-octave rainbow, as if the sky was singing!

row to trigger a visual signal to the brain. And, a trained eye can distinguish between over a million colors in side-by-side comparison tests. Hmmmm... pretty impressive afterall.

So, like, I was just reading a book on colour theory (appropriately titled "Color Theory") and one of the coolest things I learned was that there are two separate models describing the behaviour of colours when they get mixed in the real world - one scenario for coloured *light* and another for coloured *pigment* (ie. paint).

To sum up the difference: when several layers of coloured *light* are added together, the result gets lighter and lighter each time a new layer is added, whereas pigments get darker when added together. In fact, with the right combinations, a *pure white* can be achieved through the addition of light, and a *pure black* can be achieved with combinations of pigments. Theoretically.

The darkening/lightening aspect is not the only difference between the two colour models, either. They also trade their primary and secondary colour assignments.* That is, the primary light colours (red, green & blue)** are the secondary colours of the pigment model, and vice-versa (cyan, magenta & yellow).

In my mind this raised a question, which was quickly quelled upon ponderance. The question: does one of the colour models equate more directly and fully to the world of sound and music? And, if so, which one?

When lots of sounds are added together, piled on top of each other, the overall result gets louder and louder; to me, this seems more like a very bright light*** than a very dark abyss. Darkness is more akin to silence, I say. Therefore, the light model is a more effective reflection of music than the pigment model. This simple rule wants to serve amongst others as part of a complete (or at least effective) system of melding the eye-able with the ear-able. This system could be brought to life through intelligent machines which produce tones according to the colour makeup of the image collected by its camera eye, perhaps. The gadget could be installed into a pair of sunglasses or a baseball

---

* A primary colour is one which cannot be created by combining other colours within a colour model. Two primaries combine to make a secondary colour (eg. in the RGB colour space, red + blue = magenta).
** All these years, I thought the primaries were red, *yellow* and blue...??? Please refer elsewhere for more thorough and proper clarification - I'm a musician, not a freakin' scientist!
*** The common term "white noise" refers to *an equal amount of all frequencies sounding simltaneously*. Various other "noise colours" have been established throughout the noise-naming community, such as red noise, orange noise, green noise, pink noise, brown noise, black noise...

cap, outputting realtime music in response to the visual surroundings of the wearer.

Presumably, if the colour-sound system was quantized and focused to coincide with the inherent natural resonances of some important target element (such as a singer's lung cavities, an audience's average auric field frequency, a medical patient's diseased cells, a haunted room's resonant dimensions or the planet's seismic resonances, for example - assuming such things exist) then light and sound could be applied in co-operation for a synergistic healing effect, thereby rectifying the respective difficulty (vocal fatigue, tired chakras, ill organs, poltergeists, earthquakes...) I'm sure these inventions are already brewing, somewhere out there.

Alas, I digress... Returning to the transposition of visible light downward 40 octaves to coincide with musical notes, I decided to cheat a bit and reconcile the annoying octave-completion gap (of two notes width) by stretching the spectrum to encompass the Chromatic Scale, in a manner similar to buttoning up a pair of pants that's two sizes too small:

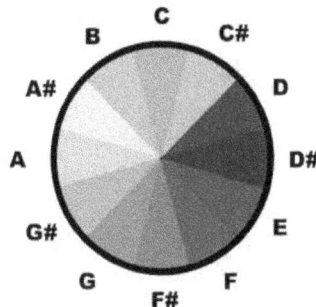

Tempered Chromatic Colour Circle

Mission accomplished - with only a small sacrifice in accuracy between E and G.* I suppose that now is as good a time as any to ponder the meaning of *chroma*.

P.S. Did you know that the ancient Egyptians used to dip themselves in vats of coloured pigment to cure their ails?

---

* Compare this diagram with the one at the bottom of p. 72 to revisit the mystery of the missing F & F#.

CHAPTER Φ

Sound Came First

Sound came first. But... was it the sound of a *chicken*, or an *egg*? This is one of many questions which I will probably never answer.*

Various cosmologies throughout the world refer to a *"sound* that started it all" when they describe the beginning of all time and creation. In one such story, the god says something like, "where's the darned light switch" and then flicked on the light. "Let's get light-headed" and - *kaboom!* There it was.** According to legend, he/she/it *spoke* the word *first*. Maybe if the god didn't have such a habit of thinking out loud (hey at least I'm not the only one who does that), the lightswitch woulda got flicked without any utterance, and we would have no sound in the universe whatsoever. That woulda totally sucked. I like music. Hell, I'd almost say I *love* music, but high-falootin' academic literature is no place for frivolous sentimentality.

For acoustical physics reasons, it seems sensible that sound *must* have existed prior to light. I explain: Since the realm of light is of higher frequency than the realm of sound, I propose that *light is an overtone produced by sound.*** The audible range of the spectrum is analogous to a musical fundamental tone, and the visible range implies an overtone series. I ask: Must not a fundamental tone exist as a pre-requisite to the existence of an associated set of harmonics? I think it must. And therefore so too is light an automatically generated by-product of sound. Maybe.

But... wait! Now I shall jump to the other side of the fence. There also exists a beast called the "subharmonic," which is generated *below* a fundamental tone. Perhaps sound is actually a subharmonic generated from light, and *light* really came first. Or, going back to the overtone concept again, perhaps both sound and light are overtones of something even lower, a more basic frequency - but, *what?*

---

*   As Bazooka Joe (who is a Mystic Master Of Space And Time and therefore ought to know about this cosmic stuff) points out, breakfast does come before lunch, so...

** According to the Hermetica, God uttered a word which calmed the chaotic waters of creation, even going so far as to call the word the "son of God" - Jesus must have been a musician, I guess. Also, the Maori legends of New Zealand say the creator *sneezed* life into the first humans. *aaaAACHOOooo!* Sound, sound, sound... perhaps all physical manifestations of energy as matter are merely expressions of musical forces. And, of course, let's not forget about the Big Bang!

*** Refer to the Glossando for detailed discussion of overtones.

Hmmm... *numbers*, perhaps?
A very short poem to celebrate the notion:

music: cosmic infrastructure

Whatever's going on here, there is definitely some kind of power in music. Music sits near the top of many people's top ten list. Many have been healed or enabled or spiritualized or driven to frenzy or confusion or rebellion or awareness or horniness by its radiance. Many of the world's spiritual practices involve sounds in their rituals. The ringing of bells has been a longtime method of dispelling thunderstorms. In ancient Egypt, a spiritual rebirthing ritual called "The Second Death" involved tapping the inside of the Great Pyramid to create special hypnotic ringing tones. Also: by yelling at the top of one's lungs for 8 or 9 years non-stop, enough energy is created to heat one full cup of coffee!

And there was that one episode of South Park which I heard about through the following email from lead singer of Sadistic Humor, Grant Hartley:

> "Did you see the "South Park" episode where Cartman and Kenny discover the "brown note" during the 4 million kid flute blow? They change the sheet music for these bully kids from New York hoping they will play the brown note (which is the exact right frequency to make a person shit their pants). Well, Yoko Ono and Kenny G and the co-ordinators get a copy and assume the music has been changed. This new music is put on the overhead projector and all 4 million elementary school kids (led by Kenny G and Yoko) play the musical score with the aforementioned "brown note" to a worldwide satellite link televised audience. Basically, the whole world shits their pants at once. The aftermath is on par with a world war or armageddon or something like that. So, the point (because Southpark is real) is that music is indeed POWERFUL....................."

Powerful indeed. And let's not forget the turntable method popularised by Cheech and Chong in the 70's.*

As yet another example of the power of music, I have a personal

---

* A contestant named Bob shared his technique on the "Let's Make A Dope Deal" game show: if you wanna see God, play Black Sabbath at 78 speed, man!

anecdote I would like to share with you. This story involves Pantera (the metal band, not the automobile).

One fine day in the mall I went for the gold in the form of a loaded hot-dog. It sure was tasty when it went down, but my guts were a real mess for about two-and-a-half days afterwards. So anyway, my food poisoning was still acting up while me & some friends were on the way to a Pantera gig in Vancouver. It looked as though my having to stop to shit projectile on the side of the road every few minutes was going to make us late for the gig. I was not impressed by this situation. When we got to the gig and had to wait outside in the lineup for awhile, the pain and the avalanche noises in my gut were so distracting that I doubted I'd be able to stick around to hear one of the best metal bands in history.

I'm glad I was wrong...

We got inside and I told my pals I gotta go sit down in the bleachers to nurse my brewing storm. I was basically immobilized by impending diarrhea. So then Pantera starts to play... the sound quality was rather large and effective and - no surprise - the band completely ruled right from the very first note. The bass frequencies were so deep and kinetic that I could feel all my guts inside start to re-arrange themselves. The innard experience was reminiscent of snakes untangling to their freedom, or poison rocks shattering into bits and settling to the bottom of a funnel, ready to be flushed into oblivion. I ran to the can, let out an explosive flourish of hellish stink, and returned to the concert hall in full force, good as new and ready to rock.

Heavy metal cured my stomach flu that day. Poetically enough, it was the "Vulgar Display Of Power" tour.

I shit you not, this really happened!

CHAPTER Φ

Has Anyone Seen The Music?

As a kid (like most kids, I guess), I spent a lot of time with crayons and various other colouring utensils. Some of those colours looked so delicious I could almost *taste* them! Sometimes I *did* taste them, but instead of raspberries or grapes or marshmallows or cherries or oranges or blueberries or chocolate or cornflowers, they always tasted like wax, paper, or wood.

Oh well. I guess sometimes imagination is more realistic than reality. However, this association between seeing and tasting still has some interesting real-world implications...

Synesthesia - blending of the senses - is a real phenomenon and seems to support the idea that, on some level, *sound and light are one and the same.* Some people really *do* taste colours with their eyes! In keeping with the tone of my quest for an objective and consistent standard of correlation between sound and light, as exemplified by the radial arrangement which I derived as shown on p. 75, my hope was that the existing body of research "out there in the world" would reveal such a system.

However, after an inquisitive trip to the library, it seems that synesthesia is a very subjective matter; different test subjects exhibited different synesthetic responses to stimuli. One person might see red while hearing F-sharp while another person sees yellow, for example. This seems to imply that there is no direct and absolute correlation of colours and musical pitches.

Dagnabbit! That's not the answer my inner scientist was looking for. My inner *artist,* however, found this affirmation of individuality encouraging.

Another area worth looking at... *chakras.* The chakras are swirling energy plexuses located at various parts of the human body. There are 7 main ones and some other ones as well. Each chakra resonates with a specific colour, in roughly the same order as the traditional "rainbow" arrangement (ROYGBIV): chakra #1 (root) is red, chakra #2 (sacral) is orange, chakra #3 (solar plexus) is yellow, chakra #4 (heart)* is green, chakra #5 (throat) is blue, chakra #6 (third-eye) is indigo and chakra #7 (crown) is white, or else takes the colour of the most dominant chakra.

I couldn't help but wonder if each chakra has its own musical note

---

* Interestingly, the Sanskrit name for the heart chakra, *Hridiyama,* means "sound without collision". Light equals sound!

as well. Some quick research has suggested a definite "yes".

My first hunch was that higher tones probably resonate with higher parts of the body such as the head and lower tones would stimulate the lower chakras.

Although - now that I think of it - *why?*

"The simpler answer is often the right one, but let's not forget the beauty of nature's *complexities*," I thought to myself before typing it into the computer. There are indoobitabably far more layers to our cosmic confetti cake than I could ever comprehend, and at this point I am merely a guitarist in writer's clothing... I might as well stay open to clues, whether simple or complex, intuitive or counterintuitive.

Maybe the pitch-to-location relation is more of a *harmonic* progression, rather than linear... systematically *offset,* like the legendary Circle of Fifths or the firing order of a gas engine. Perhaps a hopscotchy harmonic activation sequence brings special benefits to chakra work... a musical philosopher's pipe dream, or a possible reality?

In any case, I did find an interesting variety of data out there in my research travels...

Some sources described tones for the 7 chakras which match up to a C Major Scale; that is, chakra #1 resonates and responds to a C note, chakra #2 is D, chakra #3 is E, chakra #4 is F, chakra #5 is G, chakra #6 is A, and chakra #7 is B.

Other sources suggested F-sharp for the heart chakra, corresponding to a scale known as the *Lydian* mode - C Lydian rather than C major. Considering that F-sharp is directly across from C on the Circle of Fifths and that the heart chakra is the middle of the seven, C Lydian seems like a superior choice somehow. Other specifications include F* Lydian (F G A B C D E) and C D E F# G# A# B... also known as the *leading whole-tone* scale.

My hunch is that these tonal chakra relationships vary from person to person - just like musical taste - and are best discovered through practical means such as meditating to the sound of crystal bowls or moshing to an orchestra of oscilloscopes, or singing the chromatic spiral.

Assuming the C Major Scale as a model of comparison (for simplicity and convenience - no sharps or flats) and movement through the Circle of Fifths in the classic sequence F B E A D G C, the corresponding chakra order would be: 4 7 3 6 2 5 1 (heart, crown, solar plexus, third-eye, sacral, throat, root). Another alternative would be to follow the order derived through the analysis of the fraction one-over-seven, as described in Appendix 142857. In this case, the chakras

---

* F is known by some as the *tone of spirit.*

SECTION IV: DEEP LIGHT

would be fired in the order: 2 4 3 6 7 5 1 (sacral, heart, solar plexus, third-eye, crown, throat, root).

Apparently there are actually more than 7 chakras. A curious inconsistency reared its pretty little head in this area of my research endeavours. Some of my findings described the additional chakras as a finer gradation of the 7-tone spectrum into a higher number of smaller pieces situated between the 7 main ones (like the Chromatic Scale), whereas other sources indicate locations such as the palms of the hands and below the feet for chakras 8 through 18.

Rumour has it, the human chakra system is actually evolving with our world consciousness; these colours and numbers are about to be combined, split, and reshuffled! I suspect it's already happening.

And so I ponder... what role has the evolution of music been playing in the evolution of the collective human psyche? It seems to me that, even in the last couple of decades, the music of the world has been expanding its makeup to become more universal, more all-inclusive, wiser, more human, and more open. This observation could be merely a wishful projection of my own internal experiences, but I don't think so. It has become commonplace for supposedly separate genres of music to mix together in ways that were previously undreamable: opera with techno, bluegrass with rock, hip-hop with classical, country with blues... Borders are dissolving, and why not? It's the way it should be, in my opinion. And, since music plays such a huge role in so many people's lives, I wouldn't be surprised to hear of some sort of scientifically established link between music history and some pretty extreme long-term spiritual evolution of the human race.

Not like we really need science to tell us what's right...

As an additional, brief point of interest: according to the 13-Moon Calendar system,* each chakra is sealed by an electronically-charged Radial Plasma, and these are supposedly the fundamental energies from which the entire Universe is (not was) created. Very detailed and specific methods of interacting with these Seven Seals have been proposed, but the scope of this book does not allow much more than a friendly heads-up. Feel free to do some of your own research on the subject.

In addition to the mysterious chakras, we also possess *auras* - layers of coloured energy clouds. Regarding aura colours and their association with musical pitch, I've come across one source which

---

* ...the likes of which has been used by the ancient and brilliant Mayans, Incans, Druids, Lakota, Polynesians, and Egyptians; and the use of which was supported by one famous Mr. Ghandi as well as Eastman Kodak.

suggests a downward movement of frequency for an upward progression through the colour sequence red-orange-yellow-green-blue-indigo-violet - an *inverse* relation! As the light frequency goes up, the sound frequency goes down. To me, this seems like more support for one of the ideas I mentioned earlier in Section IV - the notion that sound is a *sub*harmonic generated from light. Then again, maybe the source in question had frequency mixed up with wavelength.

Yikes.

Controversy continues to abound within my mind regarding the objective truths of sound and colour. This area of my investigation remains inconclusive until further notice.

Here's a few buzzwords relating to sound and light for you to explore during your research efforts: syntonics, Cymatics, aum, Vedic mantra/ yantra, Music of the Spheres, Dinshah Ghadiali, Dr. Edwin Babbitt, luminous eggs, ultrasound, Eckankar, echolocation, dog whistles, Kirlian photography, H.A.A.R.P., the Tomatis Method...

Section V

EXPLORING THE ANCIENT SOLFEGGIO MATRIX

All but forgot for centuries, a six-tone musical scale called the Solfeggio is rumoured to possess magical healing powers rooted in mathematical resonance. I took it to task to try taking it apart and putting it back together again.

CHAPTER Φ

Introducing The Ancient Solfeggio Frequencies

The ancient Solfeggio (rhymes with arpeggio) was revealed to me in a book called *Healing Codes For The Biological Apocalypse* by Leonard G. Horowitz and Joseph Barber. The Solfeggio is a musical scale - a set of six frequencies which possess amazing yet simple mathematical resonance, and are supposed to help activate healing and spiritual "tuning." Some old forgotten* hymns were based on these tones, and some scientists even use the third tone, known as *Mi*, to repair DNA.

It seems that this amazing musical birthright has been hidden from the common person by secret societies and religions, but it is not the aim of this book to confirm, deny, defend or attack these things. I will not spend much time and space re-ierating that which is told in *Healing Codes,* only as much as is necessary to provide a background understanding for my explorations of the Solfeggio which follow.

When I was alerted to the existence of the Solfeggio and its orientation towards the healing power of music, it fit in with that dream I mentioned earlier,** and was a suitable augmentation to my previous exposure to things like meditation, shamanic drumming, chakras, didjeridoos, aum'ing, crystal bowls, psychedelic trance & ambient music, and what-have-you. These things, combined with my more previous knowledge of the emotional and visceral power of heavy metal and classical music, had already familiarized me with music's ability to influence the course of reality - to guide the flow of energy. Music isn't just a creative outlet - the harmonic, self-reflecting nature of the entire universe might *seem* like a flukey bonus for our entertainment, but sound is actually the underlying infrastructure of *everything that ever was or will be!*

The Solfeggio seemed like a very good opportunity for calculation, experimentation... *exploration.* I feel like I am a scientist, philosopher, healer, mathematician, musician, astronomer, super-hero, rock-star, wizard and authour when I dive into these mysterious calculations, even though I feel hardly qualified in any of those professions.

Devil may care, I'm writin' this book anyways!
NnnnnyYYYAAAaaaahhh!!!!

---

* or... *hidden?*
** See p. 3.

Here are the frequencies of the Solfeggio tones, expressed in hertz (cycles per second):

LA	852 Hz
SOL	741 Hz
FA	639 Hz
MI	528 Hz
RE	417 Hz
UT	396 Hz

The SOLFEGGIO FREQUENCIES

I found the frequencies very easy to memorize once I noticed the pattern - simply start with a *3* in the lower left corner of a 3-by-6 grid and count upward 3 4 5 6 7 8, wrapping under and around to 9 1 2 3 4 5 and again to 6 7 8 9 1 2 until the entire grid is filled.

Voila - the Solfeggio matrix made easy in less than 5 minutes - and doesn't stick to the pan!

The book *Healing Codes* describes some of the Solfeggio's relationship to the 8-times table and the odd alphabet here and there, as well as its encoding in a section of the *King James Bible*. I won't reiterate these aspects, but suffice it to say that I found it an inspiring read. I proceeded with my own adventures in research and experimentation, some highlights of which are told in this book.

The Solfeggio is not to be confused with the *Solfège*, which is the traditional singing exercise using the syllables *Do Re Mi Fa Sol La Ti* upwards through the 7-note Major Scale. *This* ancient Solfeggio has *6* tones, *not 7*. Some of the names of the tones are the same as in the Solfège, but the similarity ends there - the two scales sound completely different.

When the Solfeggio grid is repeated on every side of itself, some larger mystery (symetry) seems to unfold::::::::::::::

SECTION V: EXPLORING THE ANCIENT SOLFEGGIO MATRIX

Mezmerizing!

```
            5  2  8  5  2  8  5  2  8  5  2
         1  7  4  1  7  4  1  7  4  1  7  4  1  7  4
      6  3  9  6  3  9  6  3  9  6  3  9  6  3  9  6  3
   5  2  8  5  2  8  5  2  8  5  2  8  5  2  8  5  2  8  5
4  1  7  4  1  7  4  1  7  4  1  7  4  1  7  4  1  7  4  1
3  9  6  3  9  6  3  9  6  3  9  6  3  9  6  3  9  6  3
   8  5  2  8  5  2  8  5  2  8  5  2  8  5  2  8  5
7  4  1  7  4  1  7  4  1  7  4  1  7  4  1  7  4  1  7
6  3  9  6  3  9  6  3  9  6  3  9  6  3  9  6  3  9  6
5  2  8  5  2  8  5  2  8  5  2  8  5  2  8  5  2  8  5
   4  1  7  4  1  7  4  1  7  4  1  7  4  1  7  4  1  7
   9  6  3  9  6  3  9  6  3  9  6  3  9  6  3  9  6  3
      5  2 |8  5  2| 8  5  2  8  5  2  8  5  2  8  5
      4  1 |7  4  1| 7  4  1  7  4  1  7  4  1  7  4  1
         3  9 |6  3  9| 6  3  9  6  3  9  6  3  9  6  3
            2  8 |5  2  8| 5  2  8  5  2  8  5  2  8  5
               1  7 |4  1  7| 4  1  7  4  1  7  4  1  7  4
                  6 |3  9  6| 3  9  6  3  9  6  3  9  6
                  5  2  8  5  2  8  5  2  8  5  2  8  5  2  8
                     4  1  7  4  1  7  4  1  7  4  1  7  4  1  7
                        3  9  6  3  9  6  3  9  6  3  9  6  3  9  6
```

Swimmin' in a dreamy sea of numbers! Zoomin' out and out and out... the big rhythms become the little rhythms as they speed up and sound eventually turns to light...

Hmmmm... and it follows from "as above, so below" that the Solfeggio matrix must also possess a *micromesh!* Fodder for future ponderings, I suppose. Perhaps the answer will simply materialize. Questions rise, answers fall.

According to *Healing Codes*, a crucial aspect of the Solfeggio's strength is in the Pythagorean Skein pattern of the frequencies.* An interesting repeating sequence of 9 3 6 9 3 6 is revealed in the skeins:

Solfeggio Tone	Pythagorean Skein
Ut 396	3+9+6=18, 1+8=**9**
Re 417	4+1+7=12, 1+2=**3**
Mi 528	5+2+8=15, 1+5=**6**
Fa 639	6+3+9=18, 1+8=**9**
Sol 741	7+4+1=12, 1+2=**3**
La 852	8+5+2=15, 1+5=**6**

---

* Recall that the Pythagorean Skein is simply the repeated addition of a multi-digit number's digits until only one digit remains.

Try typing each of the Solfeggio frequencies using the keypad on your computer and you will experience tactile insight into the symetry of these numbers.

(Here they are again for easy reference so you don't have to keep turning back the page):

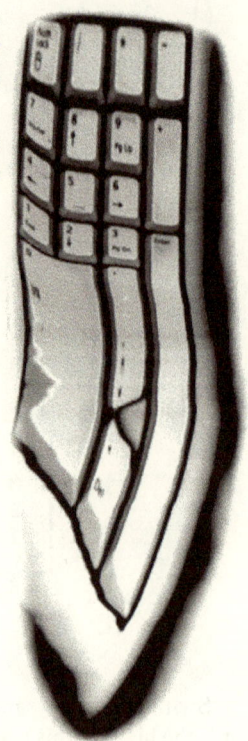

396
417
528
639
741
852

(Do it.)

396417528639741852
396417528639741852
396417528639741852
396417528639741852
396417528639741852
396417528639741852
396417528639741852
396417528639741852
396417528639741852
396417528639741852396417528639741852
396417528639741852396417528639741852
396417528639741852396417528639741852
396417528639741852396417528639741852
396417528639741852396417528639741852
396417528639741852396417528639741852
396417528639741852396417528639741852
396417528639741852396417528639741852
396417528639741852396417528639741852
396417528639741852396417528639741852
396417528639741852396417528639741852
396417528639741852396417528639741852
396417528639741852396417528639741852
396417528639741852396417528639741852
396417528639741852396417528639741852

(Okay, that's enough already.)

CHAPTER Φ

Solfeggio Octaved Skeins

My first step in exploring the Solfeggio matrix was to synthesize each tone on the computer using a wave editor. At the frequencies specified, they sounded too high for practical meditational hymning - nothing like a didjeridoo or an aum, more like R2-D2 with the stomach flu, or a super-annoying car alarm.* Too squeeky for me. Not the kind of thing *I* could dream and trance and heal to - not unless I reattuned my ears first by shapeshifting into a mouse or a fly. Stranger things have happened, I suppose.

According to my understanding of music theory, transposition of an octave is considered essentially inert,** leaving a note's harmonic function unaltered and fully intact. Don't get me wrong, the octave transposition *does* affect the effect on the listener,*** but the numerical harmonic infrastructure is oblivious (and impervious) to human taste.

So... I used the computer to lower the tones by an octave and several. This sounded much better to my ears... definitely a step in the right direction, at least.

Octave transpositions probably wouldn't affect the Pythagorean reduction numbers for each tone, I surmised. Any other transposition is not so transparent and would presumably undermine the resonant integrity of the Solfeggio system; the skein numbers would definitely be compromised in a non-octave situation, I tentatively concluded.

But... *octaves?!* Without a doubt, the Solfeggio matrix *must* be able to withstand transposition in octaves, the most benign of all musical transformations!

---

* The common person's entirely reasonable non-reaction to the impotent bleeping of car alarms - has there ever been a more ludicrous manifestation of the old adage, "never cry wolf"? Instruments of apathy, I say. There was a time when I thought noise pollution was an infeasibility (I'm part metalhead); nowadays it's more like an inevitability. Am I just getting old?

** In chemistry, there are certain gases called the *inert* gasses. They are inert specifically because they have *eight* electrons in their outer orbit. So, I see a correlation between chemistry and music here: the inert octave is the *eighth* step in a diatonic scale, and the inert gases have *eight* electrons in their outer shell.

*** The "effect" is the reason I'm about to attempt transposition in the first place, I can't deny that; the squeeky stuff grosses me out when I'm trying to zone out (unless there's lots of reverb and stereo delay on it). But this squeek-factor is probably best thrown into the pile labelled "psychology" or "physiology" rather than the "harmonic analysis" arena which claims the inertness of octaves and therefore has an overall inert effect on the discussion regarding the inertness of octave transpositions.

I mean... a G-sharp is a G-sharpis a G-sharp is a G-sharp is a G-sharp is a G-sharp is a G-sharp is a G-sharp is a G-sharp is a G-sharp is a G-sharp is a G-sharp is a G-sharp is a G-sharp is a G-sharp, *no matter which octave* of G-sharp...

Right?

So, Ut is Ut, then. No matter what octave. It simply *must* be!

Just to be sure, I thought I should check this assumption with some calculations... We all know the old "ass out of me and you" cliché.

Objective:
To determine whether transposition in octaves has any effect on the skein numbers of the six Solfeggio tones.

Hypothesis:
I predict that the skein numbers will not change when subjected to octave transpositions.

Method:
Simple division and multiplication by 2 is applied incrementally to each tone. The resulting frequency is then reduced to its Pythagorean Skein (abbr. PS) by repeatedly adding its digits until only one digit remains.

Observations:
To my utter amazement, an unexpected pattern emerged! Tones having a skein of 9 do not change with octave transpositions - just like I expected (hey... "skein of nine"... there's a song in there, I can almost hear it). *But*... the other tones *flip-flop* their skein numbers back-and-forth between 6 and 3! Check out the table on the next page to witness the incredibleness.

SOLFEGGIO OCTAVED SKEINS

Note	Frequency (Hz)	Pythagorean Skein
UT 396 Hz	3168 1584 792 396 198 99 49.5	9 9 9 9 9 9 9
RE 417 Hz	3336 1668 834 417 208.5 104.25 52.125	6 3 6 3 6 3 6
MI 528 Hz	4224 2112 1056 528 264 132 66	3 6 3 6 3 6 3
FA 639 Hz	5112 2556 1278 639 319.5 159.75 79.875	9 9 9 9 9 9 9
SOL 741 Hz	5928 2964 1482 741 370.5 185.25 92.625	6 3 6 3 6 3 6
LA 852 Hz	6816 3408 1704 852 426 213 106.5	3 6 3 6 3 6 3

Wow. I was totally wrong about my prediction - and that has *never* happened before. I guess being wrong has its benefits... afterall, I learned something new. This was one of those coffee-shop eurekas, thanks to the portability of the good ol' calculator watch; at that moment, a clear vision of flip-flopping triangles with corners labelled 3, 6, and 9 popped into my head. I couldn't help blabbing on and on about it, like there was no tomorrow. I vividly remember that moment as if it happened yesterday. Actually... I think I sort of skipped yesterday, so maybe that figure of speech doesn't really apply.

CHAPTER Φ

The Solfeggio Transposition Triangles

Flip, flop, flip, flop, flip, flop, flip, flop, flip, flop, flip, flop.

Like sandals in the sand.

3, 6, 3, 6, 3, 6, 3, 6, 3, 6, 3, 6, 3, 6, 3, 6, 3, 6, 3, 6, 3, 6, 3, 6...
9, 9, 9, 9, 9, 9, 9, 9, 9, 9, 9, 9, 9, 9, 9, 9, 9, 9, 9, 9, 9, 9, 9, 9...
6, 3, 6, 3, 6, 3, 6, 3, 6, 3, 6, 3, 6, 3, 6, 3, 6, 3, 6, 3, 6, 3...

Nine remains constant. Like a rotational axis. Or the zero-voltage axis of an oscilloscope!

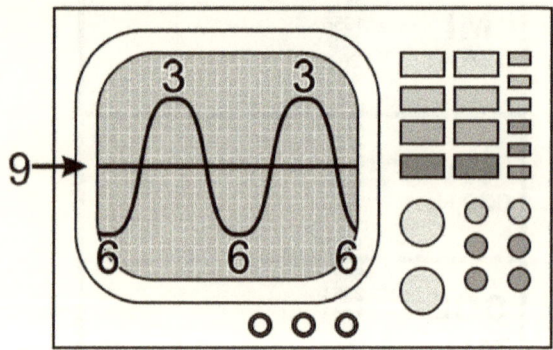

3 and 6 dancing around the Maypole of 9. Fusilli (corkscrew pasta). The twisting ladders of DNA. Twizzlers. The old *one-two*. Yin and Yang. Twisted Sister. Swingin' pendulum. Grandfather clock. The Mayan Tzolkin's Galactic Portals. Kilt of the Dervish. Center beam of the abacus. Caduceus coiling. Pluto and Charon with pockets full of posies around an invisible fulcrum. Stray cats zig-zaggin' across the back lawn.

Like so:

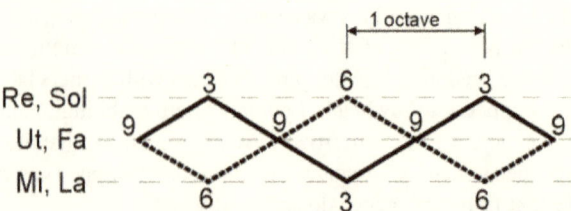

SECTION V: EXPLORING THE ANCIENT SOLFEGGIO MATRIX

And, turned sideways:

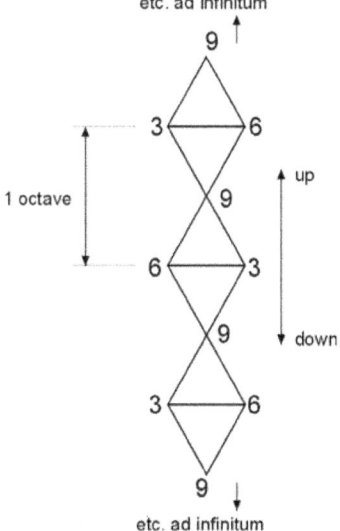

Reduced to its only two component triangles:

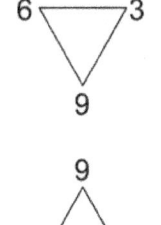

And its only two component triangle waves:

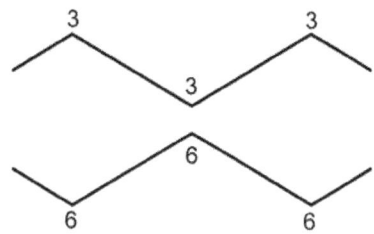

Hmmm... the 3- and 6-waves somewhat resemble "M" and "W"... What could it mean?

Men and Women? Midgets and Widgets? Moth and Wing? Me and We? Moles and Weasels? Minks and Winks? Mountains and Woolhills?

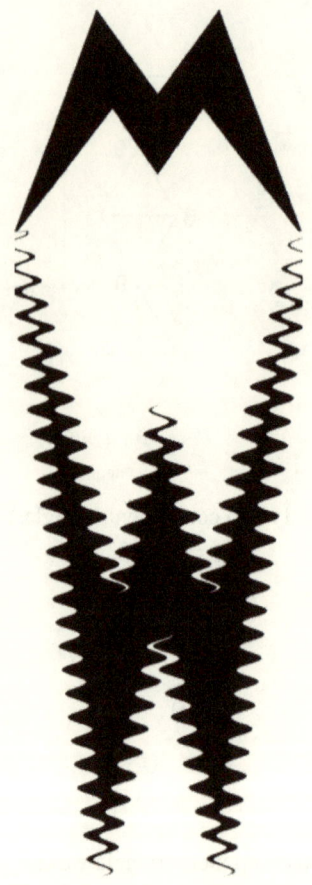

M-shaped moons reflecting on the water... what does it all mean?

It's so elegantly simple - just a bipolar pulse, really. Off and on, back and forth. The in and out, up and down of so-called "humping". Good and evil, black and white. Yada and blah, blah and yada. Does this flip-flop modulus thing have any true relevance in regards to the Solfeggio's application - and, if not, how shall I *render* it useful?

Maybe it's all a little *too* simple. I insist that we are living in a complex multiverse, so let's kick in the turboboosters and attempt Solfeggio analysis in *3-D!*

Fasten your seatbelt, here goes nothing...

SECTION V: EXPLORING THE ANCIENT SOLFEGGIO MATRIX   93

CHAPTER Φ

Solfeggio Transposition Triangle Nesting

Just one more thing, before moving on to the 3-D stuff...

Since geometry tells us that the equilateral triangle expresses a perfect musical *octave* by virtue that its outer corners lie on the perimeter of a circle which is exactly twice the size of the biggest circle that can fit inside the self-same equilateral triangle, I brainfarted the following diagram as a tidy culmination of the previous discussions regarding the now-infamous Solfeggio Transposition Triangles:

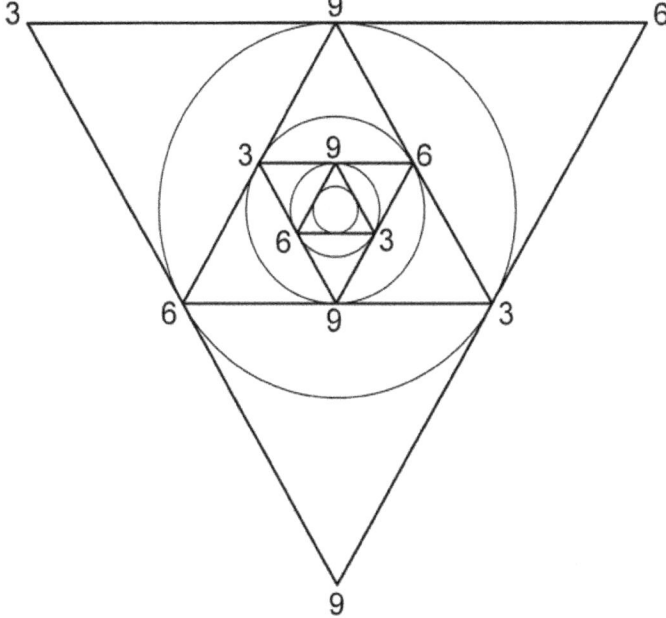

FIG. 4: SOLFEGGIO TRANSPOSITION TRIANGLE NESTING

Please ponder carefully.

Triangles seem to provide an interesting avenue for exploring the six ancient Solfeggio tones. Now let's step it up a notch and bring the trianglinesses of the ancient Solfeggio into the 3rd dimension...

CHAPTER Φ

Solfeggio Platonics

In geometry there's this thingamajigger called the *tetrahedron*.

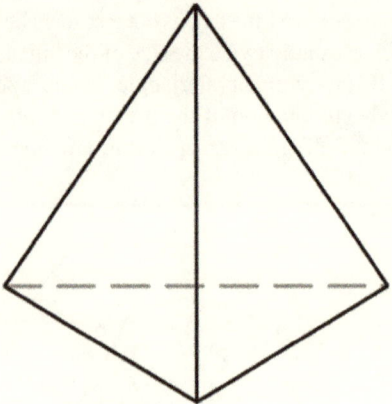

Tetrahedron

It's reminiscent of a pyramid, but the bottom is a triangle instead of a square.*

In fact, it's *completely* made of triangles... so, it only makes sense to ascend to the next layer of the perceptual onion by arranging several Solfeggio Transposition Triangles into the shape of a *Solfeggio Transposition Tetrahedron!*

Just visualize it - a numerically perfect Solfeggio matrix, spinning through the space-sound-time-consciousness continuum in a tetrahedral arrangement - *wow!* Who needs to listen to heavy metal records at the wrong speed to see God when you've got the Solfeggio Transposition Tetrahedron to spin your kundalini into voracious ataraxic bliss? Combine both approaches, and the possibilities are probably *twice* as endless!

Now that you're dizzy with meditative numerical drunkenness, check this out - another invention of mine...

The Solfeggio Transposition Octahedron!

---

* See Appendix 6, "Shine On You Crazy Tetrahedron" if you enjoy tetrahedrons.

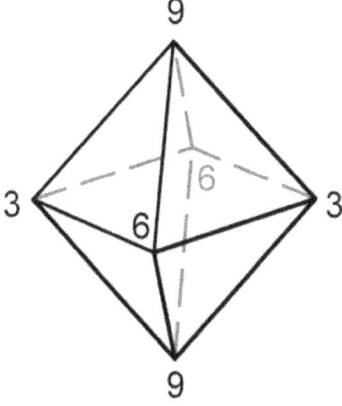

The Solfeggio Transposition Octahedron

Put a big shiny star beside the above diagram because it's one of my grandest discoveries *ever!* Cash donations accepted. Notice how all eight triangles are properly cornered by 3, 6, and 9 (who could expect anything less from a bona-fide Solfeggio Transposition Triangle), and the corner numbers all match up nicely, each vertex satisfying the requirements of four separate triangles simultaneously. Obviously, the Solfeggio Transposition Octahedron is a perfect geometrical meditation vehicle for hardcore Solfeggio enthusiasts.

Mandala supreme!

Next... the *icosahedron* (if that's its real name) is made of twenty equilateral triangles. This is what it looks like:

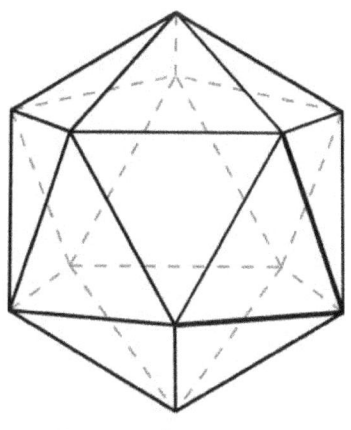

Icosahedron

Amazing, I know. I wish I invented it. Hopefully the icosahedron will accomodate the Solfeggio Transposition Triangles as beautifully as the octahedron does. But... Wait! Rewind... I forgot to put numbers on the Solfeggio Transposition Tetrahedron a few paragraphs ago...

Uh-oh... *it doesn't work!* If I start by putting numbers on the right-hand side triangle first (an arbitrary choice), the remaining vertex of the tetrahedron (the bottom left corner) doesn't know what number to be, thereby rendering the other three triangles (left, back and bottom) incomplete and confused:

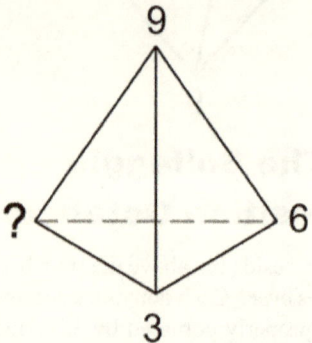

See what I mean? In the above diagram, the bottom triangle wants the question mark to be a 9, because its other two corners are 6 and 3. The left triangle wants the question mark to be a 6. The triangle at the back is crying, "Where is this 3 I've been waiting for?" The question mark is faced with an impossible 3-way dilemma.

There just seems to be no reconciliation between the Solfeggio Transposition Triangles and the classic tetrahedron. How rude! I guess tetrahedrons just aren't real, then, are they? Just kidding. That was my vertical-thinking impersonation. It's just a joke, take it easy.

Now, where was I going next? Ah yes, the icosahedron... Oh, *crap!* Same problem - some of the numbers don't fit:

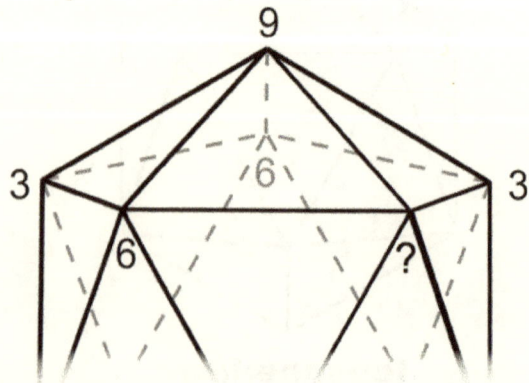

SECTION V: EXPLORING THE ANCIENT SOLFEGGIO MATRIX

The Solfeggio does not appear to be transposable across the icosahedron or the tetrahedron.

Are these the right words, grammatically speaking? Does this even make any darned sense whatsoever? I'm just trying to imagine how this would be used practically. In the case of the octahedron, where the numbers *do* work, I suppose one could transport one's consciousness from one vertex (corner) to the next, sliding along the edges to get from one point to the next. As the listener leaves the last corner and approaches the next one, the previous tone would fade away and be replaced by the new one.

Listening not only with ears, of course. Notice how I said "consciousness"? Not just hearing. The *all-pervading being* of 9, or Ut, or Fa, or whatever the case may be. The vibe of focus. The assemblage point, even. The notch in the graph - the gateway, the wormhole, the resonant spike. The frequency parameter of a hi-Q bandpass filter... The meeting point of opposing cone-tips.

And not restricted to the *edges* of the octahedron, either. If one could skate around on the *face* of Solfeggio Transposition Octahedron, the Pythagorean Skein signal would morph in a continuously variable, harmonized mix of 3, 6, and 9. In the example below, the asterix would sound mostly like 9, but with a splash of 3 and a hint of 6 in the mix as well:

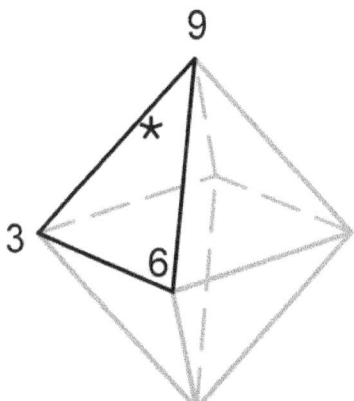

Of course, there's nothing stopping a potential candidate from diving real hardcore into the third dimension and swimming around *inside* the octahedron. Did you notice how *octave* and *octahedron* are lingually similar, synchronistically reflecting their mutual transposability in octaves through the ancient Solfeggio matrix?*

<insert spooky Twilight Zone-esque theme here>

---

* Refer to 2nd footnote on p. 87.

Here is a colour-incapacitated scan of a purple, 8-sided, octahedron-shaped die, which may be found at your typical retail comic and RPG outlets, or in the possession of the people who possess such items:

I sometimes use this one, and others like it, to write music. 8-sided and 12-sided dice are great for writing music because there are 12 notes in the Chromatic Scale and there are 8 notes in a diatonic scale if you fudge a bit by including the octave. I don't think there's any such thing as a 7-sided die.

By the way, there are other possibilities for geometric shapes comprised of triangles, but I chose the tetra-, octa-, and icosa- hedrons because they are of a specific, well-known type. They're the popular ones - and, of course, I want to be vicariously popular by including successful existing systems in my writing; henceforth their guest appearance in these pages. The shapes mentioned in this section are called "Platonic solids" (named after a famous philosopher who shall remain nameless at this point), and they have some pretty unique properties - for example, the property of self-reflection, in a somewhat similar vein to Phi itself, or a pair of mirrors facing each other. For example, joining the centres of each face of a tetrahedron will create a cube, and vice-versa. Johannes Kepler even demonstrated that the Platonic solids describe dimensional properties of our Solar System. Each of these 5 regular convex polyhedra,* as they are also called, have sides which are identical and equilateral. Each polyhedron will

---

* The 5 Platonic solids have been called the "atoms of the universe" at least as far back as ancient Greece, and each shape was equated with an element and an energy state (dodecahedron = earth, solid; icosahedron = water, liquid; octahedron = air, gas; tetrahedron = fire, plasma; cube = universe, aether). In this light, it seems that, since the octahedron is the only polyhedron that properly accomodates the Solfeggio Transposition Triangles, the 6-tone Solfeggio has a special connection with air. Perhaps the airy balance of the Solfeggio Transposition Octahedron represents mirror-like perfect duality, whereas all of the other, unresolvable "Solfeggio transposition polyhedra" represent dynamic, propulsive polarities. I wonder... does the Solfeggio Transposition Octahedron analogize with Dorian's function in The Tonal Mirror and the number *9* in The Times-Table Mirror (both described in Section III)? It does now, I guess.

also fit perfectly inside a sphere, with all corners touching the inside of the sphere. So far we've seen the tetra-, octa-, and icosa- versions of the hedrons.

There are two more I'd like to mention: the *cube** and the *dodecahedron*.

A cube is a cube is a cube - six sides, each a perfect square (unlike the typical ice-cube, which is more like a truncated pyramid). The dodecahedron looks a bit like a soccer ball - sort of; it's made of 12 pentagons, whereas a soccer ball has a bunch of pentagons and a few hexagons. Or maybe vise-versa. I've never been much of a soccer expert.

Proper vertical thinking will not allow me to deal with pentagons and squares here in the land of the Solfeggio Transposition *Triangles* because that's a different dimensional resonance that is basically imperceptible from a triangle's point of view. Forcing a Solfeggio Transposition Cube into existence would be sort of like clapping with one hand or expecting a ham radio to teach a seahorse how to speak French. It just doesn't work unless everyone's on the same wavelength.

Pentagons, however, might be well suited towards treatment of the *pentatonic* scales, which have - yup, you guessed it - 5 notes.

Pentatonic Transposition Dodecahedron analysis -wouldn't that be fun? A future project perhaps; maybe a new chapter for a future edition of this book.

In Review:

1) How to apply the oscillation of Pythagorean Skeins in the Solfeggio, induced by octave transpositions, and the underlying meaning of this oscillation, remains a mystery for now. It could imply an imbalance to be avoided, either by transposing in even numbers of octaves only, or by avoiding transposition altogether. If that were the case, then we're stuck with the high, shriekey registers, and that just doesn't seem right. To me, the modulation seems to imply an inherent vital dynamic pulsation, as in a heartbeat. The question remains of whether or not transpositions of the Solfeggio by octave increments shall be considered *inert* or *active* -

---

* a.k.a. the hexahedron - represents the number 6, and relates to the planet Saturn somehow.**

** Many moons after writing the above footnote, I read something so amazing that I suspect a hoax: at the North pole of planet Saturn, a very distinct and mysterious hexagon in the clouds has been photographed by investigative probe droids! Hoaxagon? Hoaxahedron?

musically, spiritually, mathematically, functionally.
2) The octahedron nicely accommodates my metaphoric visualization/contraption/invention known as the Solfeggio Transposition Triangles, in effect manifesting a 3-D mandala known as the Solfeggio Transposition Octahedron.

Now... where was I going with this?

All this geometrical philosophilisation stuff is pretty entertaining unto itself, but it should ultimately serve the development of a practical usage scheme for the Solfeggio.

The quest for a system continues...

CHAPTER Φ

A Nasty Surprise

My next strategy in attempting to understand the Solfeggio tones would be to layer the actual sounds of the tones, so that they all played at once. I was hoping (okay, I admit it - maybe I was even *assuming*) that it would sound perfect. I thought to myself something like, "When I play the Solfeggio tones all together at once, I will become instantly enlightened; all the planets will line up and I will be able to levitate and cure all the ailments of everyone I meet with the blink of an eye. I shall become... Superman! King Midas. Übermensch. Enlightened. Mega-buddha-riffic, that's me! Nobel prize, the cover of *Popular Science...*"

So I fastened my seatbelt and played all of the tones at once...

(drum roll, please...)

It sounded terrible!

Yuck!

Absolutely horrisonant. After recovering from my disappointment, I raised some questions. First came doubt in the validity of the tones; maybe some of the numerical values put forth in *Healing Codes* are incorrect, despite the seeming perfection of their derivation and their appearance in ancient texts of supposed wisdom. After all, the Solfeggio spans more than an octave, which *does* seem a little suspicious from a music theory point of view; as far as I can recall, most every scale I've ever encountered as a musician completes itself within an octave and echoes its spelling throughout all other octaves in both directions, re-establishing the root as a brand new seed point for expansion at each new octave.

Like a couple of mirrors facing each other, ever-repeating.

Like the Golden Ratio, self-replicating. Self-reflecting.

Perpetual motion machines playing ping-pong.

Silly spirals.

Some scales vary according to whether the melody is climbing the ladder or coming down the ladder, but the one-octave containment limit and ripitition in the spelling are still maintained... not so with the Solfeggio.

I should point out here that sounding *every* tone of *any scale* all at once would probably sound a little overwhelming and un-musical to the average listener. To hear what I mean, play lots of notes on the piano at once by leaning your arms across the keyboard. It sounds like a bit of a mess, eh? Nevertheless, all of those tones within the mush *do* create chords with each other, and hints of melody abound within the chromatic soup, indistinct and unintelligible as they may be. The Solfeggio matrix, on the other hand, seems to contain some sourness of incompatibilty which prohibits an all-inclusive mixture even more strongly than the well-armed piano does.

However, I decided to assume that the Solfeggio frequencies are correct, because disproving them is a quest less inspiring than actuating them. I concluded that there must be a special way to arrange the tones to render each of them fully useful and at least somewhat harmonically versatile with the other tones - some specific set of rules governing their combination, in a similar vein to the chord theories of the more common standard musical scales.

So, my next step was to test different combinations of tones through trial and error. I did this in a preliminary, experimental and qualitative manner, which is to say that I made generalized observations only, and didn't write down any of the specifics.

I absorbed a few impressions, so to speak. What I found was that, while a certain pair of tones sounded perfectly fine together, and another pair might also sound good together, they sound absolutely terrible when combined into a three- or foursome, or if they traded partners. Some of the tones seemed to be more monogamous and/or finicky than others. Kinky stuff, these freaky frequencies!

Also, certain tones (such as Re) simply do not seem to play well with others... black sheep tones in the Solfeggio?

Are the Solfeggio tones a cliquey bunch who hang out in small packs and avoid the other freaquey freaks? Unbefitting, I say.

Shocking! The drama unfolds... Will all six Solfeggio tones have a chance to let their vibrations shine? *I think so!*

CHAPTER Φ

Massive Solfeggio

As much as I fiddled around with various combinations of tones in groups of two and up to five, a straightforward correlation to the familiar chromatic system refused to emerge, thereby demanding the creation-slash-discovery of a much more organized, quantified system.

But first, to take advantage of my "green perspective" on the situation, I attempted to create a good piece of meditation music based on a subjective, non-systematic approach. Not *completely* lacking structure (I can hardly ignore two decades of musical conditioning), but without the advantage of having yet investigated the internal math of the Solfeggio frequencies.

In consideration of the ultra-serious approach which I do profess not to have chosen, I thought that attempting a composition may seem like frivolity at this stage in the game - why not wait until the tones are properly understood before actuating them, rather than making an amateurish mess? Why would I let myself become the creator of a sonic dog's breakfast by putting the cart before the horse?

Because a fresh perspective has its advantages, that's why. Some of the best art is made by novices, in my opinion. A blank slate provides space for the intuition to roam unprejudiced by the conditioned intellect.

So, I went ahead and created a 21-minute piece of music called "Massive Solfeggio." I selectively avoided most of the tense combinations in attempt to provide what I hoped to be a gentle-but-giant meditative space. At the time of this writing I, and a few sporadic volunteers, have taken the plunge into "Massive Solfeggio," with some quite positive results. Some have found it a bit too dramatic, others find it relaxing. I've enjoyed auming to it, and sometimes listen while writing or reading or doing some quasi-yoga or playing drums.

However, I wouldn't recommend the didjeridoo in conjunction with the Solfeggio if you intend to reap the benefits of the pure, undisturbed mathematical resonance of the Solfeggio, because there is a high potential of harmonic conflict between the Solfeggio and any fixed-pitch instrument which is not specifically tuned to the Solfeggio. Auming, on the other hand, is not what I would refer to as a fixed pitch because the person singing has the ability to adjust the pitch of their voice so that it resonates in harmony or unison with the Solfeggio if they wish. The same goes for trombone and slide guitar, whereas instruments such as didjeridoos, flutes, saxophones, and marimbas have more stone-like tonal collections which are more likely to cause a chaotic structural collapse within the presumedly fussy, purist Solfeggio matrix.

On the other hand, in the interest of creative freedom and good ol' participatory involvement, don't let authours of books tell you what not to do.

CHAPTER Φ

Solfeggio Intervals

Anyways, at this point in the game it seemed like a good idea to start really inventing a workable and meaningful Solfeggio system. Or, rather, *finding* the understanding of the Solfeggio's inherent inner sensibilities. If such sensibilities even exist; a healthy bit of skepticism should help objectivity, I thought - perhaps the Solfeggio matrix is only a silly math oddity, a propaganda scam without any practical musical rhyme or reason... snake oil. Any conclusions I could hope to find are only giving part of the picture, I assume. Every deep-sea explorer finds their own unique treasure.

My next step was to quantify the relations between the tones. In simple math terms, this means finding the ratio between every possible pair of Solfeggio tones. In terms of music theory, this means searching for correspondences with chromatic interval names such as the perfect fifth, minor third, etcetera.

A table would do the best job of presenting this kind of information, I decided. Using such a *Solfeggio Interval Table*, a composer such as myself (or anyone else) could then create Solfeggio music which capitalizes on the system's inner symetry more effectively, thoroughly and easily.

This may be premature since you ain't had a chance to see the table yet (it wouldn't fit on this page), but I'll just proceed to share some of the observations I made while filling out *the table*...

The first calculation, 417 Hz divided by 396 Hz, gave an interesting result - a *repeating* decimal of 1.053030303030303030303030... In the context of Just Temperament, this interval of Re:Ut approximates a minor second, which is 25:24, or 1.041666666666666... Therefore Re:Ut is 18.78 cents higher than a minor second. That's according to this formula, where n and p are the two frequencies being compared:

$$\text{cents} = \log(n/p) \times 3986.3137$$

Considering that 100 cents equals a semitone, 18.78 cents deviation isn't exactly what I would call a very close and accurate match to a minor second, so it seems to me that Re:Ut might be its own special animal, living in a mysterious kingdom unbeknownst to Just Temperament..

Turn the page to unveil the Solfeggio Interval Table, whose reputation ought to precede itself by now.

THE SOLFEGGIO INTERVAL TABLE

		Ut 396 Hz	Re 417 Hz	Mi 528 Hz	Fa 639 Hz	Sol 741 Hz	La 852 Hz
Ut 396 Hz	Solfeggio Interval (decimal) →	Ut:Ut = 1.0̄ = 1:1	Re:Ut = 1.05303̄	Mi:Ut = 1.3̄ = 4:3	Fa:Ut = 1.6136̄ ≅ Φ = 71:44	Sol:Ut = 1.8712̄	La:Ut = 2.15̄ = 71:33
	Nearest Just Interval Ratio →	Unison = 1:1 = 1.0̄	m2 = 25:24 = 1.0416̄	P4 = 4:3 = 1.3̄	m6 = 8:5 = 1.6	M7 = 15:8 = 1.875	m2+8ve = 25:12 = 2.083̄
	Solfeggio Deviation From Just →	0 (perfect)	+18.78 cents	0 (perfect)	+14.69 cents	-3.50 cents	+55.70 cents
Re 417 Hz	Solfeggio Interval (decimal) →		Re:Re = 1.0̄ = 1:1	Mi:Re = 1.26618705...	Fa:Re = 1.53237410...	Sol:Re = 1.77697841...	La:Re = 2.04316546...
	Nearest Just Interval Ratio →		Unison = 1:1 = 1.0̄	M3 = 5:4 = 1.25	P5 = 3:2 = 1.5	m7 = 9:5 = 1.8	m2+8ve = 25:12 = 2.083̄
	Solfeggio Deviation From Just →		0 (perfect)	+22.27 cents	+36.96 cents	-22.28 cents	-33.71 cents
Mi 528 Hz	Solfeggio Interval (decimal) →			Mi:Mi = 1.0̄ = 1:1	Fa:Mi = 1.2102272...	Sol:Mi = 1.403409̄	La:Mi = 1.6136̄ ≅ Φ = 71:44
	Nearest Just Interval Ratio →			Unison = 1:1 = 1.0̄	m3 = 6:5 = 1.25	Tritone = 45:32 = 1.40625	m6 = 8:5 = 1.6
	Solfeggio Deviation From Just →			0 (perfect)	+14.69 cents	-3.5 cents	+14.69 cents
Fa 639 Hz	Solfeggio Interval (decimal) →				Fa:Fa = 1.0̄ = 1:1	Sol:Fa = 1.15962441...	La:Fa = 1.3̄ = 4:3
	Nearest Just Interval Ratio →				Unison = 1:1 = 1.0̄	M2 = 9:8 = 1.125	P4 = 4:3 = 1.3̄
	Solfeggio Deviation From Just →				0 (perfect)	+52.48 cents	0 (perfect)
Sol 741 Hz	Solfeggio Interval (decimal) →					Sol:Sol = 1.0̄ = 1:1	La:Sol = 1.1497 97570850202429
	Nearest Just Interval Ratio →					Unison = 1:1 = 1.0̄	M2 = 9:8 = 1.125
	Solfeggio Deviation From Just →					0 (perfect)	+37.74 cents
La 852 Hz	Solfeggio Interval (decimal) →						La:La = 1.0̄ = 1:1
	Nearest Just Interval Ratio →						Unison = 1:1 = 1.0̄
	Solfeggio Deviation From Just →						0 (perfect)

My next calculation was a total treasure find! Mi-over-Ut equals an exact perfect fourth ratio of 4:3! Awesome. The Solfeggio-skeptic in me was suddenly assuaged. But I had no idea what was lurking around the corner of the next calculation...

SECTION V: EXPLORING THE ANCIENT SOLFEGGIO MATRIX   107

Proceeding rightwise across the top row of the table, Fa-over-Ut is next in line:

639 / 396 = 1.6136363636363636363636363636363636...

Wow! This number is *very* close to the Golden Number (1.61803398874...) with the added bonus that it's a *repeating* decimal, a veritable horn-of-plenty full of oh-so-notorious 6's & 3's! Plus, it reduces to a rather interesting-looking fraction of 71/44.* Need I say more? With numbers like these popping up everywhere, my Solfeggio enthusiasm waxed significantly.

Until the widest interval came along, that is... dividing La-over-Ut rained a mystified wave of confusion and disappointment all over my happy little parade. Darn!

852 / 396 = 2.1515151515151515151515151515151515...

NNNOOOOOO!!!!!!!!! It *can't* be! I thought this Solfeggio thing was supposed to be a musical *scale*. No scale that I've ever encountered contains an interval with a ratio of *more than an octave!* A scale typically completes itself *within* an octave, such said octave being a perfect 2:1 ratio above the root, above which the scale re-iterates itself. Intervals wider than an octave are typically given names like ninth, eleventh, or thirteenth, and are mere duplicates of those which occur within the octave (9=2, 11=4, 13=6).

If the last Solfeggio interval is *more* than an octave, then what's supposed to happen as we ascend into the next iteration of the matrix? Should the frequency series "backslide" down to the nearest octave to begin the next go-round at Ut's second octave, as in Ut=396 Re=417 Mi=528 Fa=639 Sol=741 La=852 Ut=***792***, thereby suffering some serious slippage in the drivetrain? Or should Ut's next frequency jump way up to the *next* octave all of a sudden, as in Ut=396 Re=417 Mi=528 Fa=639 Sol=741 La=852 Ut=***1584***? That seems equally senseless - such a harsh jump-up with a disproportionately wide range of unused frequency spectrum seems rather un-scale-ish, in my perception. What a waste of wavelength! Then again, skipping an almost-octave in this manner *would* circumvent the skein-alteration dilemma inherent in octave transpositions described in the "Solfeggio Octaved Skeins" Chapter, since the Solfeggio would only exist in every second octave.

Drawn as a spiral model, with radial lines echoing true octaves, the Solfeggio appears to trip all over itself like a breakdancing pretzel**

---

* See "Meet You On The Dark Side Of The Mean" in Section II.
** Did you hear the one about the pretzel walking down the street? It was assaulted...

with two left feet:

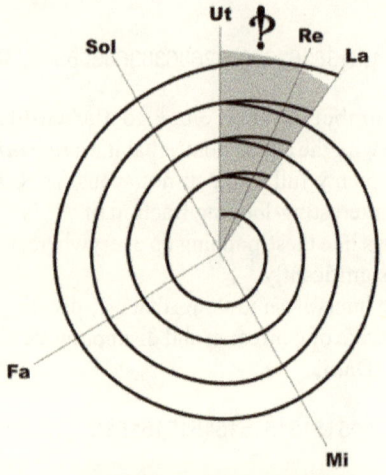

**The Solfeggio Spiral:
Jagged And Stubborn
Tail-Swallowing Overlap**

... like a set of nested lower-case sigmas! Re-calibrating the diagram with the 6 tones evenly spaced on the radial front and octaves represented by concentric circles, the anomaly begins to remind me of an *off-side* in soccer:

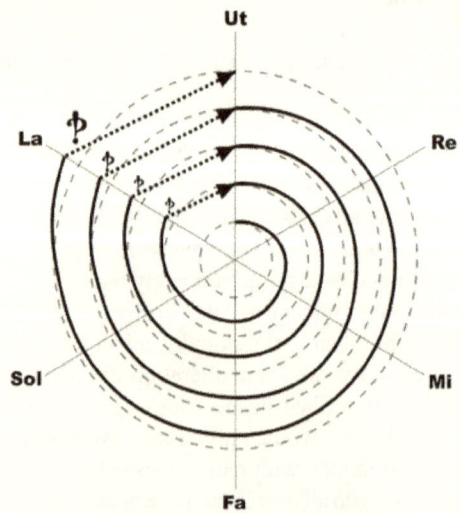

**The Solfeggio Spiral:
Off-Side Wrinkle Conundrum**

I mean football, I mean soccer, I mean - uhhhhh...

Perhaps the Solfeggio has its own unique concept of octave, something other than the simple logic of the 2:1 ratio...

Uhhh, NO. That's just crazy! I insist that "octave" is synonymous with "double." The establishment of a fudging device such as a "Solfeggian comma" may be in order.*

My last 20 years of musical training did not prepare me for this. But, then, a new question is every bit as good as an answer, I suppose - maybe better. So the mystery goes... another intriguing find on the treasure hunt through the sands of math and music. Another chunk of chaos gets thrown to the bottom of the in-box for future processing.

Continuing with the bright side, La-over-Ut's troublesome 852:396 reduces to 71:33, which is indubitably reminiscent of the previously-hailed "golden" 71:44 which shows up elsewhere on the table. Sweeeeet! Maybe, by some quantum or metaphorical link, the overlap is meant to pick up the slack where our human visual frequency range falls short of an octave.** Comparing the 33 and 44 also yields the power chord ratio - as one would expect, considering that La:Fa is 4:3.

A few more observations from the Solfeggio Interval Table...

Sol-over-Ut is a slightly-flat major seventh. This is full of compositional potential because major sevenths gravitate intensely up towards the octave (root) of a scale, creating strong melody. La-over-Ut (mentioned on the previous page) is an octave plus a minor second, but not very in tune with Re-over-Ut, which is also a minor second. This caused me to wonder if the frequency specified for La in *Healing Codes* was incorrect; however, this suspicion quickly waned upon recognition that La-over-Mi is 71:44, a.k.a. 1.61363636... This is exactly the same ratio as Fa:Ut - a *double* occurence of Phi! What a frigging awesome discovery.*** Being a nerd is *so* much fun!

Observe the very long repeating decimal in La-over-Sol. Also noteworthy is the rather accurate tritone combination of Sol and Mi. Although I happen to *like* tritones as much as I like any other interval, they're evil according to the status quo so I'll probably end up leaving them out of the healing meditation music which I intend to create - just to stay out of trouble and not accidentally possess anyone of Satan.

---

* In the same vein as the Pythagorean comma encountered on p. 41.
** Refer back to the complaints on p. 73.
*** This incredible mirrored consistency echoed in my mind the non-Solfeggio phenomenon known as the *tetrachord*. In music theory, the tetrachord is a specific sequence of four notes, with the distances between them arranged in this very specific sequence: whole-step,whole-step, half-step (abbreviated W W H). By stacking another tetrachord on top of the first, starting a whole-step higher than where the first one left off, the entire complete Major Scale is created (W W H W W W H).

CHAPTER Φ

The Quasi-Palindromic Triangular Solfeggio Sequence

Armed with the Solfeggio Interval Table, I could now proceed to brew up a more proper composition.

Objective:
Compose a harmonic progression which explores the inner symetry of the Solfeggio matrix.

Parameters:
1) The sequence should favour consonance over dissonance (it should be pleasant, in other words... this music is meant for meditation, afterall).
2) Include all six tones in the sequence.
3) The progression will favour harmony over melody. Because* melody is compositionally too easy to get away with and doesn't reveal the math as obviously, whereas harmony is a less arbitrary challenge and will really help make the Solfeggio's structure more apparent.
4) Ut is the root.
5) The progression will occur in 3 phases, with the same number of steps in each phase.

Procedure:
Phase 1 of 3
1) Start with Ut. The root is always a good place to start.
2) Add Mi to Ut; the perfect fourth is perfectly harmonious.
3) Add Fa to Ut and Mi. This creates a three-tone harmony which includes a minor third between Mi and Fa and the Golden Number between Ut and Fa.
4) Add La. Now Fa and La harmonize a perfect fourth, while La harmonizes Phi with Mi as 71:44.** Also, Ut and La create a somewhat tense-sounding but numerically-interesting ratio of 71:33. Fa with Ut

---
* Never start a sentence with "because." Why? Don't ask me why.
** This 4-piece harmonic structure reminds me of the tetrachord structure mentioned on the previous page: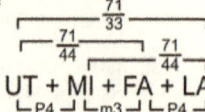

SECTION V: EXPLORING THE ANCIENT SOLFEGGIO MATRIX

also rings the Golden Number, and the perfect fourth is also present between Ut and Mi. This 4-part Solfeggio chord is a *harmonic powerhouse!*
5) Subtract Ut from the chord, because of its uncomfortable dissonance with La.
6) Drop Mi from the mix, leaving Fa and La to sing a perfect fourth together.
7) Drop Fa. Now La stands alone as a point of repose, having travelled a long way from Ut and ready to serve as a new beginning for Phase 2.

Phase 2 of 3
Whereas Phase 1 featured a build-up and teardown of dense multi-voice harmony, Phase 2 will endeavour to combine smaller intervals - those in the neighbourhood of a major or minor second - in small, blustery clusters of two or three tones.

1) Add Sol to La. This harmony approximates a major second.
2) Drop La and add Re to Sol. The interval of a minor seventh is created - equivalent to a major second.
3) Drop Sol and add Ut to Re. The interval of a minor second is the result.
4) Drop Re and add Sol to Ut. The major seventh is the resulting harmony, which is equivalent to yet another minor second.
5) Add Fa to Ut and Sol. The Golden Number appears between Ut and Fa. Ut and Sol make a minor second while Sol and Fa sing the major second. This is a "Solfeggio cluster" if there ever was one.
6) Remove Ut. Sol and Fa continue the major second.
7) Drop Sol. Fa stands alone as the culmination of Phase 2.

Phase 3 of 3
Phase 3 is basically the same as Phase 1, except backwards. Partial palindrominity ensues therefrom.
1) Add La to Fa; a perfect fourth is the result.
2) Drop Fa, leaving only La.
3) Add Fa again. (Sorry about the weird offset, it's the only way I could get this thing to work.)
4) Add Mi to Fa and La.
5) Add Ut to Mi, Fa and La.
6) Subtract La from the mix, leaving Ut, Mi and Fa.
7) Drop Fa, leaving Ut and Mi to sing a perfect fourth.

Beyond this point, dropping Mi from the mix would return the sequence to the beginning of Phase 1, with Ut standing alone as the root. The round trip is complete.

Results:
The diagram below shows the sequence in graphical format:

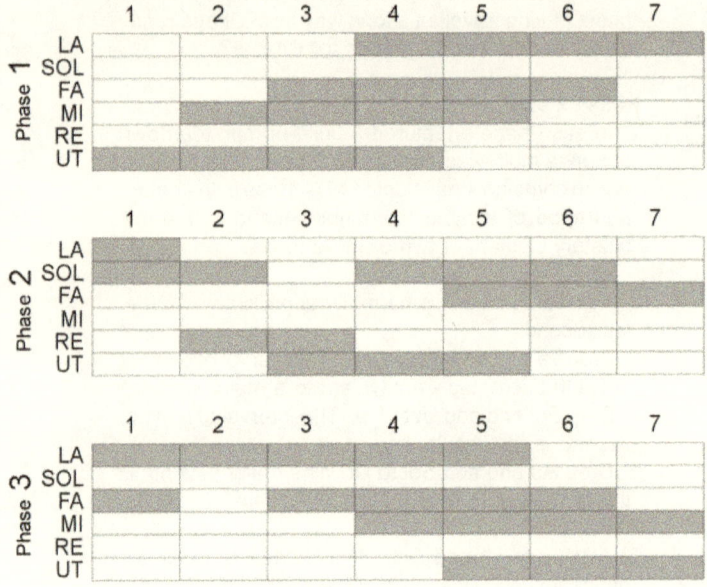

This Quasi-Palindromic Triangular Solfeggio Sequence has already been rendered (par moi) into hearable music.

CHAPTER Φ

Solfeggio Fretboard

Gee, I almost forgot to solve the most obvious question: where are the Solfeggio tones found on a guitar? Mere approximations, of course, since the Solfeggio doesn't jive with Equal Temperament...

```
Ut  = 396 Hz     ~ G       = 392 Hz
Re  = 417 Hz     ~ A-flat  = 415.3 Hz
Mi  = 528 Hz     ~ C       = 523.25 Hz
Fa  = 639        ~ E-flat  = 622.26
Sol = 741        ~ F-sharp = 739.98
La  = 852        ~ A-flat  = 830.6
```

A-flat occurs twice and is therefore self-absorbedly redundant in the context of multi-octave scale useage; Equal Temperament turns the Solfeggio into a snake that swallows its own tail! With its piece of tail missing, the Solfeggio becomes a *five*-note scale instead of six.

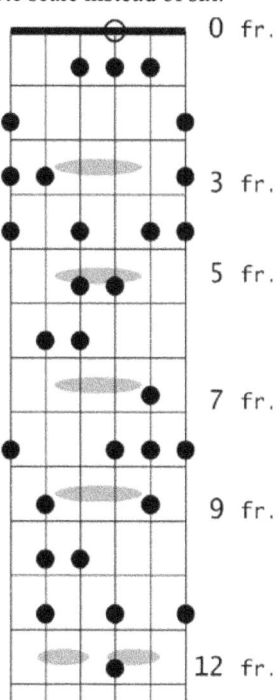

This changes *everything*, doesn't it? What will the finicky skein symetry have to say about this?* Maybe it's not such a bad thing... Pentagons and therefore the realm of the soccer ball become available through this metamorphosis - and five-sided geometry vehemently implies the Golden Ratio! There's got to be some interesting parallels worth exploring in there *somehow*.

Moving forward, I present to you the Solfeggio fretboard diagram in all its glory. I chose not to circle all the "G" notes as is customary in guitar-land, since I find Ut's domain as the root of the Solfeggio to be rather questionable.

For example... Presuming the Major Scale's worthiness as a universal standard, *Re* might be a worthy candidate for root, since Re-Mi-Fa-Ut spells an A-flat major seventh chord, perfectly manifesting the Major Scale.

...Solfeggio modalism, anyone?

---
* Refer back to p. 85 for a memory refresher.

CHAPTER Φ

Circular Graphs Seem To Reveal A Lot

Arranging the Solfeggio tones around the outside of a circle, like this... 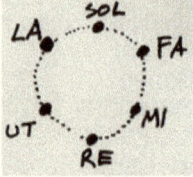 ...with their skein reductions arranged likewise...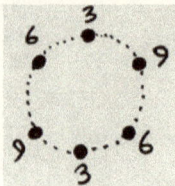

...some interesting patterns emerge...

An instant parallel with the world of colour theory comes forth, as matching numbers are to be found across from each other in the wheel, echoing the *complementary colours* concept. As described in the "Deep Light" Section, the light colour model establishes the primary triad as red-green-blue, while its secondary triplet is cyan-magenta-yellow.*

Each one of these makes a triangle.

Applying the light colour model to the Solfeggio matrix, we have the "Primary Solfeggio Triad" Ut-Mi-Sol. If Ut is considered the musical root of the entire Solfeggio, then Fa should be considered the root of the secondary trinity, since it is complementary to Ut in its skein of 9; therefore, the "Secondary Solfeggio Triad" is Fa-La-Re.**

Each triangle spells 9-6-3 with its Pythagorean Skeins, which, according to the Solfeggio Transposition Triangles (probably

---

* The pigment model reverses the distinction between primary and secondary, and shall be ignored in this section pursuant to the conclusion established on p. 74 regarding the light model's superiority in its sound-likeness.

** 2 down-quarks plus 1 up-quark equals 1 neutron; 2 up-quarks plus 1 down-quark equals 1 proton. Therefore, just for shits and giggles, the upward-pointing "Primary Solfeggio Triad" could be renamed the *Solfeggio Neutron Triad,* and the downward-pointing "Secondary Solfeggio Triad" could be called the *Solfeggio Proton Triad.* Or, maybe my triangles are upside-down... why would I start with Ut in the lower right, and why progress the labelling *counter*-clockwise? I don't have an answer - I guess it was an arbitrary choice.

irrelevant to this discussion but nevertheless I share my clutching at the straw...) seems to echo a one-octave phase shift from the original order of 9-3-6.

The 9-skein complementary pair Ut:Fa sings a 71:44 Phi harmony.
The 6-skein complementary pair La:Mi sings a 71:44 Phi harmony.
The 3-skein complementary pair Sol:Re squawks out a harmonically irreverent 1.7769784172...

The complementary pairs also share digits in each other's frequency specifications:

$$396 \ \& \ 639$$
$$417 \ \& \ 741$$
$$528 \ \& \ 852$$

The primary and secondary Solfeggio triads each contain a musical perfect fourth, represented as the horizontal 6-9 legs...

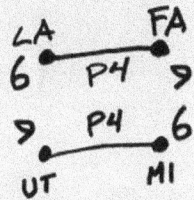

...poised facing each other from opposite sides of the circle - ready to generate a dynamic field of harmonic interplay not unlike the poles of a spinning magnet.

Connecting 6-9 pairs in the vertical direction, the primary and secondary triangles become invisibly linked by the mysterious 71:33 sharp-octave on one side and a vague Fa:Mi minor third on the other:

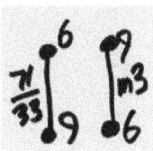

In attempt to justify the relevance of the two vertical 6-9's and not knowing what better approach to follow, I divided 71:33 by Fa:Mi. The eye-candy-ish result: 1.7777777777777... also known as 16:9 or one-and-seven-ninths. Flipping this upside down into 9:16 and raising it by an octave, we get 9:8 - a perfectly tuned major second according to Just Temperament and the Pythagorean scale of re-iterated fifths.

The 6-3 pairs do their own imitation of symetry... La:Re is roughly an octave (36.96 cents sharp - almost a minor ninth) and Sol:Mi is a tritone. The observation-of-questionable-relevance here is that tritone and the octave are on opposite sides of the Circle of Fifths and the Chromatic Circle; arguably the most dissonant pair out of the entire 12-tone chromatic system, thereby establishing yet another polar field - rhomboidal this time:

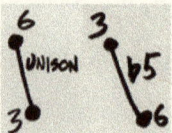

Serendipities abound in the Solfeggio colour wheel! Whoah - brainfart alert - these aforediscussed colour correspondences ought to be applied to the Solfeggio Transposition Octahedron, made available to the masses on the side of cereal boxes or the back page of the mad Solfeggio magazine in a cut-and-fold, "insert tab A into slot B" format!

P.S. Reversing the method described in the "Deep Light" Section, I transposed each of the Solfeggio tones *up* 40 octaves, into the range of visible light, to determine a "truer," alternative colour correspondence.
Behold:

```
Ut = 688.53nm = red
Re = 653.86nm = orange
Mi = 516.40nm = green
Fa = 426.70nm = violet
Sol = 367.96nm (or x2 = 735.92nm) = outside visible range*
La = 320.02nm x2 = 640.04nm = orange
```

---

* Unless you're a bird or some other sort of tetrachromat.

SECTION V: EXPLORING THE ANCIENT SOLFEGGIO MATRIX

CHAPTER Φ

Solfeggio Poetry Via Numerology

"A Poem For MiLa"

Tweet Tweet
Tweek The Beet
Then Be Zen

The Ten-Hen Web
We Knew
Bent The Tent

Then Went Between
Between Thee
Between The Knee

Tweek
Tweek
(Tee Hee)

"Solfeggio Phi"
(ver. 1.0)

I Foil Or I Fool
I Curl Or I Coil
I Roll
I Furl
I Riff
I Fill
I
Roi

"The Longest Shortest Solfeggio Poem"

Crucifix
Between
Pyjamas

"Solfeggio The Fourth"
(ver. 1.0)

Quench The Quitter
Bench The Chicken
Liquid Thinker
Crinkle Brink
Quote Her Trike
Tequila Queen
Treble Kettle

"For ReFa"

Raid My Food Jar -
Rip Crap!
Drop Flaccid, Jailfrog.
Sad, I Riff Spacy.
Afraid, I Fry A Jaggy Mass.
Add A Lumpy Lug.
Oi!
A Pricy Folio.
I'm A Saggy-Ass Vulgar Ox,
You Frilly Fox.
I Say I Jam A Clam,
A Savvy Pam,
A Coily Gypsy Ass...
Crimp My Gamma Ray!

Section VI

APPENDICES

*According to appendicitis, the appendix
is a useless appendage.
I beg to differ!*

SECTION VI: APPENDICES

Appendix 272659.6521

Converting Light Wavelength To Sound Frequency

Objective:
Devise a convenient formula for converting colour wavelength to sound frequency.

Parameters:
For the purposes of this example let's assume a light wavelength of 420 nanometers (= 420 x 10^{-9} m)
Speed of light = 299792458 meters per second (in a vacuum)
Frequency is expressed in cycles per second, a.k.a. Hertz (Hz)

Procedure:
1) Divide the speed of light by wavelength to obtain frequency:

299792458 / 4.2x10^{-7} = 713791566666666.666 Hz

2) This is an extremely high frequency, way beyond the human limitation of about 20,000 Hz. Therefore transpose down several octaves until it enters audible range. As it turns out, this is about a 40-octave transposition.

713791566666666.666 Hz / 2^{40} = 649.189648 Hz

3) Distill into a convenient conversion formula:

where
w = wavelength of coloured light in nm
f = sound frequency of coloured light after 40-octave downward transposition

f = {299792458 / (w x 10^{-9})} x (1 / 2^{40})
1/f = {(w x 10^{-9}) x 2^{40}} / 299792458
1/f = w / 272659.65218248
f = 272659.6521 / w

Results:

f = 272659.6521 / w

Appendix 5

Hertz Donut

A joke I learned in Elementary School:

Go up to someone - someone who's fair game, of course, and ask them if they would like a *hertz donut*.
They may act confused at first but, if you can get a "yes" out of them,* punch them in the arm and say, "Hurts, don't it?"

Appendix A

The Infamous Australian Treble Clef Dilemma

Rumour has it, water spiraling down a drain rotates in one direction in the northern hemisphere and the other direction below the equator.**
Perhaps it stands to reason that the Australian treble clef ought to be drawn backwards as well? Discuss amongst yourself.

While you're at it, you may wish to consider the apparent similarity between a seahorse and an earlobe!

---

* This shouldn't be too difficult; *everyone* loves doughnuts. Just don't try playing this trick on the cops!
** !emag emas eht yalp senolcyC

Appendix GGG

A Symmetrical Joke

Q: What do you get when you cross a seahorse with a starfish?
A: A galloping gas giant!

The aforetold joke is funny *because:*
  a) both initial elements are themselves dualities of contradiction.
     (ie. There are no horses in the sea, and a fish is not a star.)
and because
  b) a star is a *giant* ball of *gas*
and because
  c) horses *gallop*
and because
  d) three g-words in a row is great *alliteration,* thereby providing unstoppable poetic integrity to the joke.

Appendix 100

Golden Nice

I've devised an application for the Golden Ratio in the area of human relations. It is a modification of the technique popularly known as "meeting someone halfway." Instead of meeting someone halfway on an issue, which is exactly 50%, throw in an extra 11.80339887%. Take a little extra step of diplomacy. Not too much, just the right amount: eleven-point-eight-zero-three-three-nine-eight-eight-seven-four-nine-eight-nine-four-eight-four-five-eight-two-four-seven-four-five-eight-four-three-two-seven-eight-two-six percent, and it can be comprised of goodwill or patience or good listening or a compromise or whatever. This extra little push can really help. If both (or all) parties do this, there will be more than enough peace to go around.
Geometrical diplomacy* rules!

---

* "Sincere diplomacy is no more possible than dry water or wooden iron."
   - Joseph Stalin

Appendix Φ

The Golden Ratio With Lots Of Decimal Places

1.6180339887498948482045868343656381177203091798057 6
2862135448622705260462818902449707207204189391137 4
8475408807538689175212663386222353693179318006076 6
7263544333890865959395829056383226613199282902678 8
0675208766892501711696207032221043216269548626296 3
1361443814975870122034080588795445474924618569536 4
8644492410443207713449470495658467885098743394422 1
2544877066478091588460749988712400765217057517978 8
3416625624940758906970400028121042762177111777805 3
1531714101170466659914669798731761356006708748071 0
1317952368942752194843530567830022878569978297783 4
7845878228911097625003026961561700250464338243776 4
8610283831268330372429267526311653392473167111211 5
8818638513316203840052221657912866752946549068113 1
7159934323597349498509040947621322298101726107059 6
1164562990981629055520852479035240602017279974717 5
3427775927786256194320827505131218156285512224809 3
9471234145170223735805772786160086883829523045926 4
7878017889921990270776903895321968198615143780314 9
9741106926088674296226757560523172777520353613936 2
1076738937645560606059216589466759551900400555908 9
5022953094231248235521221241544400647034056573479 7
6639723949499465845788730396230903750339938562102 4
2369025138680414577995698122445747178034173126453 2
2041639723213404444948730231541767689375210306873 7
8803441700939544096279558986787232095124268935573 0
9704509595684401755519881921802064052905518934947 5
9260073485228210108819464454422231889131929468962 2
0023014437702699230078030852611807545192887705021 0
9684249362713592518760777884665836150238913493333 1
2231053392321362431926372891067050339982265263556
2090297986424727597725655086154875435748264718141 4
5127000602389016207773224499435308899909501680328 1
1219432048196438767586331479857191139781539780747 6
1507722117508269458639320456520989698555678141069 6
8372884058746103378105444390943683583581381131168 9
9385557697548414914453415091295407005019477548616 3
0754226417293946803673198058618339183285991303960 7
2014455950449779212076124785645916160837059498786 0
0697018940988640076443617093341727091914336501371 5
7660114803814306262380514321173481510055901345610 1
1800790506381421527093085880928757034505078081 4...

Appendix .

A Word On Decimal Places

A quick word on decimal places:

Ever heard of an anagram? That's when you rearrange the letters of a word or phrase to create a new word or phrase. For example, the phrase *"a decimal point"* can be re-arranged into *"I'm a dot in place."* Pretty cool, huh? It's so awesome that I'm surprised I didn't invent it myself.

This next one's really impressive:

"Astronomical observations" becomes *"to scan a visible star or moon."*

Fricking brilliant! Whoever discovered that one first deserves a shiny new telescope for sure.

Since we're on word games and this is sort of a book about music, here's a pun written by punmaster Alan F.G. Lewis:

"Why piccolo profession like music that's full of viol practices, confirmed lyres, old fiddles, and bass desires? For the Lute, of course."

Ha ha ha ha ha ha ha. The *loot*, get it?

Appendix 4

Number-Boiling

When I was a kid of approximately a decade's decay,* I invented this mental game involving numbers and letters:

    1) Pick a number, any number.
        (eg. 23)
    2) Spell it.
        (eg. twenty three)
    3) How many letters in the spelling of the number?
        (eg. t-w-e-n-t-y-t-h-r-e-e has 11 letters)
    4) Spell the new number and count its letters.
        (eg. e-l-e-v-e-n has 6 letters)
    5) Repeat the process for as long as possible.
        (eg. six has 3 letters, three has 5 letters, five has 4 letters, four has 4 letters, four has 4 letters... 4 4 4 4....)
    6) Notice how, no matter what number you pick as your first number, *4* is the ultimate destination. If you don't believe me, go ahead and try it for yourself with a bunch of other numbers - big numbers, small numbers, decimals or no, it doesn't matter. Four is always the final result. The game gets pretty boring once it's been played a few times, so I guess it's more of an observation than a game.

A really wild thing happened about, oh, less than five years ago (over 20 years later). Someone actually told me the same idea, and that she came up with the same thing on her own when she was a kid. Intense synchronicity, I'd say. Not the most common topic of conversation - probably just trying to steal my fire.**

As a kid I also clearly and frustratedly visualized the "unstoppable force against an immovable object" paradox, long before being exposed to the axiom. Lots of time to think when you're a kid, I guess. By the way, indulging and refining your musical experiences (consumption as well as creation, if they are really divided) is reputed to increase visualization capabilities, I think. I might have read that somewhere; maybe it's true. Anyway, back to the number-boiling.

My more recently-conceived extrapolation of this idea, an area

---

* We're born dying, right?
** See "moebius strip" in the Glossando.

SECTION VI: APPENDICES   125

for further observation, is as follows:
  This game of number-boiling might be language-dependent; what sort of numbers will appear in other languages?
  I think I'll start with French since it's the closest thing I have to a second language...

French
  1) I choose 23 (vingt-trois).
  2) "Vingt-trois" has 10 (dix) letters.
  3) "Dix" has 3 (trois) letters.
  4) "Trois" has 5 (cinq) letters.
  5) "Cinq" has 4 (quatre) letters.
  6) "Quatre" has 6 (six) letters.
  7) "Six" has 3 (trois) letters.
  8) "Trois" has 5 (cinq) letters.
  9) "Cinq" has 4 (quatre) letters.
  10) "Quatre" has 6 (six) letters.
  11) "Six" has 3 (trois) letters.
  12) "Trois" has 5 (cinq) letters.
  13) "Cinq" has 4 (quatre) letters.
  14) "Quatre" has 6 (six) letters.
  15) The pattern so far is (10 3 5 4 6 3 5 4 6 3 5 4 6...)
  16) Notice how French 23 does not boil down to a single digit, but seems to repeat a four-number cycle of 3 5 4 6.

  What a sursprise! Surprises are not welcome in science; therefore I shall choose a new number to test the French number-boiling:

  1) I choose 17 (dix-sept).
  2) "Dix-sept" has 7 (sept) letters.
  3) "Sept" has 4 (quatre) letters.
  4) "Quatre" has 6 (six) letters.
  5) "Six" has 3 (trois) letters.
  6) "Trois" has 5 (cinq) letters.
  7) "Cinq" has 4 (quatre) letters.
  8) It seems to me that, in French number-boiling, as soon as 3 (trois) is encountered, the pattern of 3 5 4 6 settles into a   pretty secure and stable orbit. Based on this admittedly preliminary attempt, my hunch is that *all* French numbers eventually boil down to the repeated pattern 3 5 4 6.

  Feel free to pick up the ball and do a more thorough investigation.
  Next up... Spanish! I don't know Spanish (not like I'm a French language guru either... or English, for that matter) but that's what Spanish-English dictionaries are for...

Spanish
1) I choose 23 (veintitres) again.
2) "Veintitres" has 10 (diez) letters.
3) "Diez" has 4 (cuatro) letters.
4) "Cuatro" has 6 (seis) letters.
5) "Seis" has 4 (cuatro) letters.
6) Spanish 23 quickly terminates in a repeating 4   6 pattern.

Another Spanish number:

1) I choose 60 (sesenta).
2) "Sesenta" has 7 (siete) letters.
3) "Siete" has 5 (cinco) letters.
4) "Cinco" also has 5 (cinco) letters.
5) Spanish 60 boils down to 5.

Weird! Two different results this time.
Another Spanish number:

1) I choose 19 (diecinueve).
2) "Diecineuve" has 10 (diez) letters.
3) "Diez" has 4 (cuatro) letters.
4) "Cuatro" has 6 (seis) letters.
5) Another 4   6... On this admittedly incomplete investigation, it seems that Spanish is a double-boiler. Also curious is how one boil is encompassed by the other (5 is between 4 and 6 on the numberline).

I wonder what sort of results would come from other languages... Ukrainian, Polish, Secwepemc, Japanese, Sanskrit, Chinese (not to mention Enochian, Loxian, Babytalk, Rodian, Klingon...) if the number-boils were found for each language, the data could probably somehow be applied musically in the form of a song or some kind of harmonic structure.

Perhaps the number-boil goes so far as to define a resonant frequency for any given language - the math of each tongue. With over six thousand languages currently being spoken in the world, that makes for quite a symphony - a boiling sea of numbers!

Appendix 4

The Circle Of Fourths

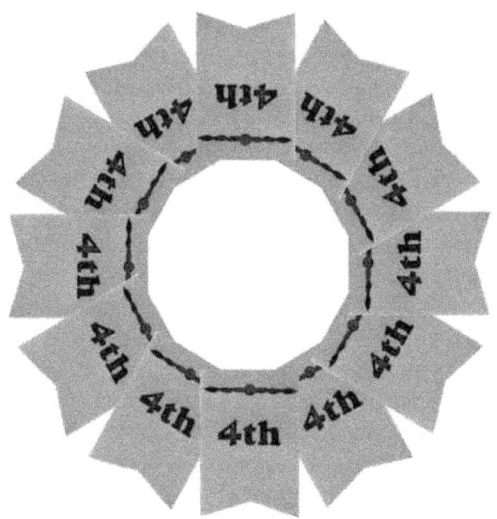

Fig. 4: Found object (sports ribbon) arranged into a circle of fourths

The *Circle of Fourths* is the almost-equal opposite of the Circle of Fifths. That is, a counter-clockwise progression as opposed to clockwise - an entirely arbitrary arrangement of course, since music is the same no matter which side of the mirror you're standing on. There is definitely a difference in the *sound* of each direction of movement.

Mathematically, a musical perfect fourth is 4:3. Notice how this is essentially the inverse of 3:2, which defines the perfect fifth. Except that the fraction contains a 4 instead of a 2; this non-discrepancy represents the octave transposition which is necessary only for the convenience of restricting interval discussion to the upward direction. This avoids the hassle and confusion of trying to explain how "a fifth down" means something completely different from "the fifth, below."

"The fifth, below" actually means the same thing as "down a fourth." Conversely, "the fourth, below" actually means "down a fifth." This is starting to sound a lot like football.

I mean, soccer... no, *football*.

Appendix 4:44:44 04/04/04

Eight Fours In A Row

Luckily I saw it coming, since it won't happen again for another thousand years. On the 4th day of the 4th month of the 4th year of this millennium, I grabbed my camera just in the nick of time. Holding the camera and my calculator watch both in the same hand in front of the bathroom mirror, I snapped. At exactly 44 seconds after 4:44 in the afternoon, I pressed the button.

I suppose I was actually late, since 4 p.m. is actually the 16th hour of the day, not the 4th. Maybe I ought to switch over to using the 24-hour time system to avoid this sort of tardiness in the future.

Then again, if I re-write the rules and decide that any current day actually starts at noon instead of midnight, my photographic adventure becomes 100% legit!

Which isn't such a bad idea - most people are already falling asleep by around midnight anyways, so they kind of miss out on the customary transition into the next day. What a ripoff, I say! I thought the changing of the guard was supposed to be a public spectacle, not a big secret. Maybe the changing of the date should occur at noon instead of midnight. After all, the word "noon" sounds like "new," and "midnight" contains "mid."

Anyways... now, 13 days later, perusing the developed prints over a pint, I reap the fuzzy rewards of my wishful syzygy. Despite my diligent efforts at finding the perfect focus, composition and, of course, *timing*... my synchronistic masterpiece came back totally blurry, the numbers completely indiscernable:

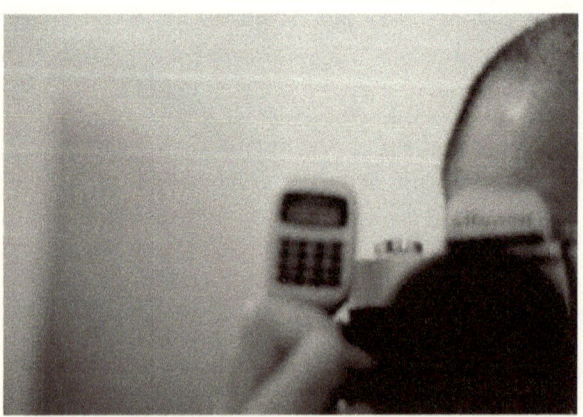

Dangnabbit!

Apparently, mirrors have a funny way of affecting depth perception. And - a funny thing... as I lament at my photographic failure over this teary-eyed beer while writing my sad story on this napkin, I overhear another patron of the pub lament his unsuccessful Keno lottery attempt with the words, "I guess the numbers just didn't line up..."

Spooky coincidence... he wasn't close enough to read what I was writing, but he *may* have been reading my mind.

FIG. 4: TEAR-STAINED NAPKIN

Timing, indeed! Now if I could just figure out all these darn-fangled buttons on this here machinery, who knows what might happen! Nothing of great consequence, probably. This book would have been better 'cause you'd be able to see all the fours perfectly lined up on my calculator watch but I guess that's about it - no big loss, better luck next time.

And, returning to the topic of the clock, my opinion of the moment is that the clock should actually start at sunrise. So 12 a.m., the official beginning of the day, should be situated where 6 a.m. currently resides. This is a clockwise shift of 6 hours compared to the current standard. Six over 24 is 0.25. Our watches are all out of whack by 25%! In terms of circles that means we're all 90 degrees out of phase with reality.

@!#?@!

Appendix 0.55555

5wimming

On the 5th day of the 5th month of this 5th year of this millenium,* I swam. Not in my usual lazy backstroking way, no. I really *went for it*, crawling all the way to the other end of the pool as fast as I could, trying not to resort to dogpaddling or backstroking or treading water. Before initiating my heroic attempt, I contemplated the sign which stated the length of the pool as 25 metres. I asked myself, "Twenty-five metres? Is that, like, one full lap, or halfway?" From one end to the other, I decided - half a lap. "That makes it simple, since that's about as far as I can swim anyways..."

"Okay, I'm ready." So, I get into position and wait for the blue hand to reach the top of the clock, and I'm off like a yellow submarine. Even though I tried really hard, I didn't quite make it to the other end of the pool without having to resort to the dogpaddle. But I kept going. And, at the precise moment I pantingly tagged the far side wall, the blue hand reached the 45-second mark. I'm sure that doesn't exactly measure up to the Olympics but I'm not competing so, like, no sweat, eh!**

A few minutes later, after catching my breath, another attempt... I tried even harder this time. Yeah, yeah, I know - "...do or do not, that is not the question." Anyways, I failed to beat my first attempt, but I didn't fall short, either. Again, exactly 45 seconds! This surprised me. What are the chances of that? I didn't rig the results by slowing my pace for a second at the end, or anything like that.

With such amazingly consistent data, I felt compelled to calculate my speed. Speed equals distance per time...

Twenty-five metres divided by 45 seconds reduces to 5 over 9.

One ninth equals 0.1111111 repeater, multiplied by five equals...

(drum roll, please)

On this 5th day of the 5th month of the 5th year, my swimming speed was exactly 0.55555555555555555555... metres per second.

That's a lot of fives, eh?

I was even wearing my pentagram T-shirt that day. Not while swimming, of course, but...

I'm a star. Five-pointed!***

---

* Hmmm... I wonder what happened 5,000 years ago?
** If a swimmer sweats underwater, can anybody tell?
*** Standing with feet a few feet apart and arms level out to the fingers, we humanoids resemble thereof.

SECTION VI: APPENDICES 131

Appendix 5

F.O.S.F.
(Found On Sidewalk Five)

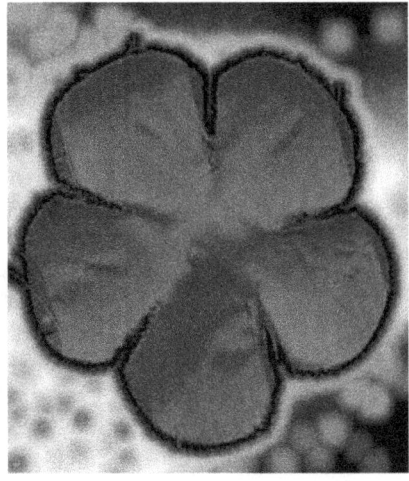

FIG. 4: ANOTHER FOUND OBJECT

I actually found two of these in one day (hey - another octave!), in two different parts of the city. One of them I kept, the other I tossed back onto the sidewalk where I found it. I figured it had a destiny elsewhere.

Appendix 14

F.O.R.F.
(Found On Road Fourteen)

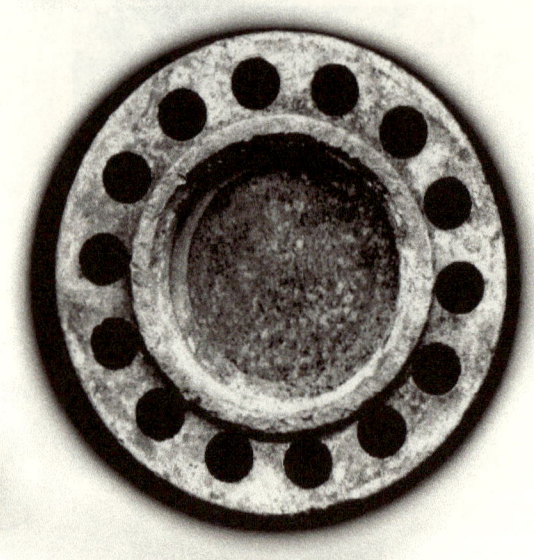

Fig. 4: Another found object

Found objects can be a lot of fun for an artist or anyone who's easily entertained.

A question rises: if the above object had some other number in its symetry, would I still have been intrigued enough to carry it all the way home and scan it into the computer, or would I have simply tossed it into the ditch? I recently acquired the knowledge (at a résumé writing workshop, of all places) that odd numbers of objects (as well as odd-numbered symetry, presumably) are somehow "more pleasing to the eye" than even numbers.

The F.O.R.F. shown above has 14 holes, which doesn't even satisfy that requirement unless you consider it an octave of 7. Yet, I still dig it. Hmmm, almost a pun there, except I didn't have to dig it out of the ground, it was just sitting there with a few chunkoids of loose dirt situated tenderly upon its cute, round face.

Appendix 142857

1/7 As A Decimal

It's relatively easy to memorize the decimal equivalents of many common fractions. 1/2=0.5, 1/4=0.25, 1/3=0.33333..., 2/3=0.66666..., 3/2=1.5, 1/9=0.11111..., 4/9=0.44444..., 5/9=0.55555...,* etcetera. One that never made it into my head back in school days all those years ago was the fraction of *one-seventh*. I finally figured it out last May. The logic of 1/7 is so bizarre yet unexpectedly simple.

Using a calculator (or abacus)...

One divided by 7 equals 0.142857142857142857142857... notice the repeating sequence *1 4 2 8 5 7*. Tough to remember? Not so tough when you break it down into three groups of two digits: 14'28'57.

The first 2 digits are a multiple of 7 - *double,*** in fact! 2x7=14.

The next 2 digits are *double double* 7; 2x2x7=28.

The last 2 digits are *double double double* 7, plus 1; 2x2x2x7+1=57.

Zero point 142857 repeater. No problem! Unforgettable from now on. You're welcome very much. But... *why* the convenient and hintingly relevant doubling of 7 in the pattern? And, why the "plus one" non-sequitur?

Wait - there's more to tell... what if you want to know the decimal equivalent of 2/7 or 3/7 or 4/7 or 5/7 or 6/7? No problem. The same six-digit sequence, 1 4 2 8 5 7, repeats itself, starting with a different digit each time:

1/7 starts with 1: 0.142857'142857'142857'142857'...
2/7 starts with 2: 0.285714'285714'285714'285714'...
3/7 starts with 4: 0.428571'428571'428571'428571'...
4/7 starts with 5: 0.571428'571428'571428'571428'...
5/7 starts with 7: 0.714285'714285'714285'714285'...
6/7 starts with 8: 0.857142'857142'857142'857142'...

Notice that they appear in the order of their size (1 2 4 5 7 8) and you've got all those fractions memorized, too!***

I derived a musical application for this number system. The next step is to render it into a song - the system is ready for implementation...

---

* Refer to Appendix 0.55555.

** The word *double* has developed a habit of inspiring sweet notions of the octave within my mind.

*** Notice anything else? Who's *missing* in this 7-divisor system? **3, 6 and 9!** Refer to Appendix 6693, of course.

Including the zero, there are 7 digits; 7 is easily connected to the musical Major Scale or any other of the umpteen heptatonics floating around out there.

So... using the C Major Scale,* a correspondence aligns and a decoder key is established...

1/7	=	0.	1	4	2	8	5	7...
C Major Scale	=	C	D	E	F	G	A	B
chord	=	C	Dm	Em	F	G	Am	Bdim
interval	=	R	M2	M3	P4	P5	M6	M7

So, what of it? What now? I want this stuff to transform into music. I suppose I could make up some arbitrary sequence out of the digits or something, but - alas - no contrivance is necessary, since increasing portions of 7 already provide the "first digit" sequence described on the previous page: 1 2 4 5 7 8.

A chord progression is about to emerge.

Since the sequence ends in 8 - which corresponds with the perfect fifth and chord number 5 according to the above decoder key - and since chord number 5 is customarily the most effective chord for ending a sequence en route back to the root, the root shall follow the dominant in this chord progression. According to the decoder key, the root is represented by 0. Put the *0* at the end of the size-matic sequence and it is done: 1 2 4 5 7 8 0.

Using the decoder key, this sequence is rendered as the chord progression Dm, F, Em, Am, Bdim, G7, C...**

"The Ballad of One-Over-Seven."

---

* C major could easily be substituted with some other key for a different tonality.
** To play as a melody or bassline instead of chords, simply play the notes in sequence while ignoring the "m's" and "7's".

Appendix Φ

The First 34 Fibonacci Numbers

Fib (0) = 0
Fib (1) = 1
Fib (2) = 1
Fib (3) = 2
Fib (4) = 3
Fib (5) = 5
Fib (6) = 8
Fib (7) = 13
Fib (8) = 21
Fib (9) = 34
Fib (10) = 55
Fib (11) = 89
Fib (12) = 144
Fib (13) = 233
Fib (14) = 377
Fib (15) = 610
Fib (16) = 987
Fib (17) = 1597
Fib (18) = 2584
Fib (19) = 4181
Fib (20) = 6765
Fib (21) = 10946
Fib (22) = 17711
Fib (23) = 28657
Fib (24) = 46368
Fib (25) = 75025
Fib (26) = 121393
Fib (27) = 196418
Fib (28) = 317811
Fib (29) = 514229
Fib (30) = 832040
Fib (31) = 1346269
Fib (32) = 2178309
Fib (33) = 3524578
Fib (34) = 5702887

Appendix EOOEOOEOOEOO...

The Fibonacci Waltz

As a curious experiment without expectation, I decided to graphically separate the Fibonacci Sequence into odd and even numbers. The surprising result is a 3-beat rhythm:

Fib#	EVEN	ODD
0	X	
1		X
1		X
2	X	
3		X
5		X
8	X	
13		X
21		X
34	X	
55		X
89		X
144	X	

etcetera ad infinitum...

Appendix 21

Fibonacci 21: A Musical Exploration

Objective:
Compose a piece of music based on the famous Fibonacci Sequence.

0 1 1 2 3 5 8 13 21

Parameters:

Project Time Allotment: Spend less than one day composing, recording and mixing this piece. It's a quick gratification kind of affair... a numerisonic one-night stand, so to speak.

Tonal Center: C, because it's the lowest note on my synthesizer and I'm going to need room for upward expansion. This will ultimately be transposed to the Solfeggio tone of Mi, since it's already so close to C.

Rhythm: Because of the Project Time Allotment and my hardware sequencer's limitation of 32 beats per bar maximum, I choose 21 as the upper limit of Fibonacci development, thereby establishing a repeating 21-beat rhythm. In musical terms this translates to a time signature of 21/4.

Tempo: I'll use a starting tempo of 136, for some reason that has now escaped me. After transposition from C (261.63 Hz) up to Mi (264 Hz), this tempo becomes approximately 136.2229 BPM.

Melodic Movement: Integrate the Circle of Fifths into the process somehow.

Procedure:
1) **Construct a Fibonacci Matrix**, in the form of a graph, in which successive levels of fragmentation into Fibonacci subcomponents (via reverse movement through the sequence) are shown as stacked layers of horizontal rectangles, each rectangle having a width of a specific Fibonacci number. Each horizontal layer of rectangles will be played by its

own musical instrument, and each rectangle will be manifested as a musical note which has a rhythmic duration corresponding to its width.

2) **Choose A Direction For Time:** Time moves from left to right on the graph.

3) **Establish A Melodic Development Algorithm:**

- Horizontal movement (ie. through time) from one "rectangle" to the next on any given level effects a melodic movement of one musical perfect fifth, thereby propagating a Circle of Fifths on each level of instrumentation.

- A direction needs to be chosen for the propagation through the Circle of Fifths - should all instruments start at "C" and move away from it, or should "C" be the ending point? This question echoes the concept of entropy, a term which customarily pertains to pursuits such as chemistry. A point of convergence shall be established on the timeline, where all instruments play the tonal center of the composition simultaneously. Think of "all instruments starting on the same note and then racing away from the tonal center through the Circle of Fifths." This point of unison, which could actually be placed either at the start or the end of the cycle, reminds me of planetary conjunction. Proliferation at multiple, non-uniform rates through the Circle of Fifths generates further and further removal from the tonal center, approximating chaos or increased entropy. For this composition I've chosen a system of *decreasing* entropy (increasing order), which equates to placing the syzygy of unison at the right-hand edge of the matrix. Thus, the musical progression begins with more chaos and ends with more order - a pattern reminiscent of the formation of crystals or the freezing of water. This choice of decreasing rather than increasing entropy was partly inspired by a desire to create a piece of peace in a world that is rampant with decay - our universe is unravelling like a dismantled cassette tape tossed out on the side of the highway, and this piece of music is trying to clean up some of the mess by pulling the other way in a tug-of-war versus entropy. Compositionally this translates into "all instruments starting at various scattered locations throughout the

Circle of Fifths and gravitating home to the tonal center". Racing at a fairly slow tempo of approximately 136 beats per minute, mind you.

4) **Other Compositional Considerations:**

- The smallest unit of fragmentation, 1, is assigned the convenient and sensible rhythmic value of one beat.
- Notice how I've avoided further fragmentation of 1's into their Fibonacci subcomponents of 0 and 1 because this process would quickly become a neverending vortex, which I hereby deem impractical for the scope of this project. It could be manifested musically *somehow*, I'm sure. A future enhancement of this project, perhaps. Maybe it's as simple as sounding a backwards cymbal at every occurence of the appearance of "1" in the non-reduced version and sending the audio signal through a pitch shifting device set at the following parameters:

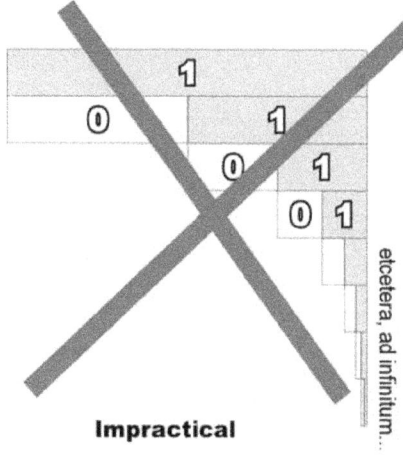
Impractical

    +8 semitones
    62 millisecond delay
    61.803% feedback
    61.803% mix

- Since a measly 21 beats create a very short piece of music insufficient for continual enjoyment beyond 9.24913 seconds or so, I've chosen to repeat the sequence for increased sustenance. This gives the divine pattern a chance to "sink in" to the listener's psyche, as well as affording some space for dynamic fluctuation of the structure, by fading in & out the various instruments. The drawback of this looping strategy - and it's a big one - is that any ripitition whatsoever is not a true representation of the Fibonacci Sequence, which actually evolves in a non-repeating pattern. A truer, more ideal version of Fibonacci music would fill its timeline with a larger Fibonacci number and (d)evolve without ripitition. Another project for the future, perhaps. The mission is yours if you choose to accept it.

Results:

I rendered a simple graphical representation of the procedure (shown below) as well as a traditional staff notation version (on the next page). Audible rendering has also ensued.

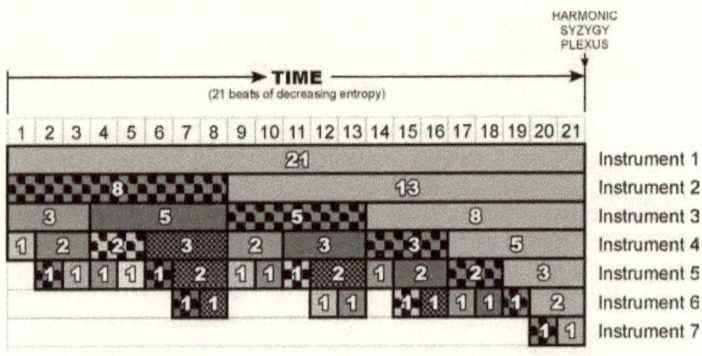

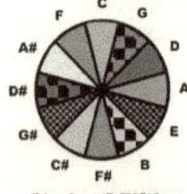

Circle of Fifths

Fibonacci 21
construction plan for a musical composition

SECTION VI: APPENDICES

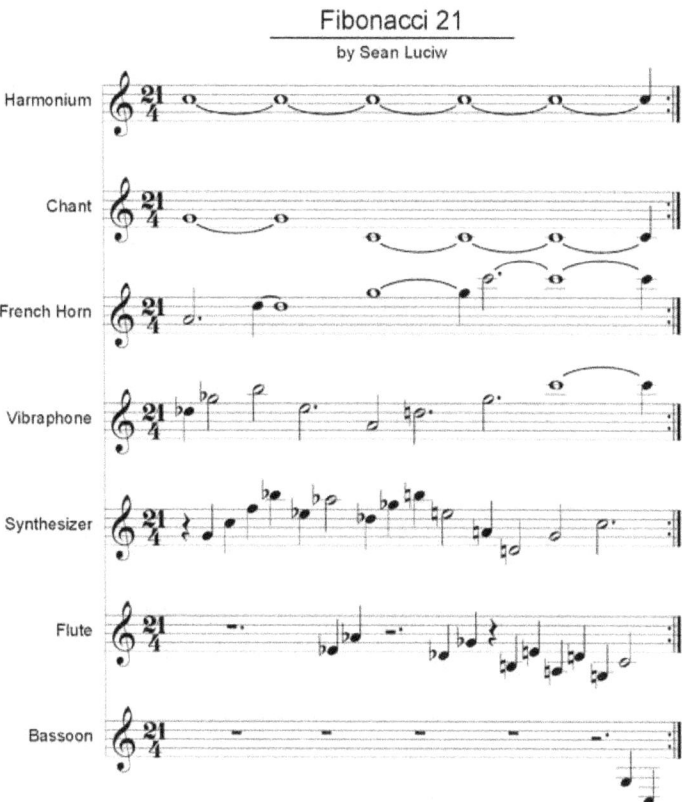

Appendix 21

Bottle Cap

This beer bottle cap has 21 spikey things around the edge of it. Once again, proof that numbers are responsible for *everything* in the whole entire Universe!

Fig. 4: Another "found" object
(okay fine I admit I paid for it)

Appendix 6693

Clock Pie

One fine day as I sat in the food court waiting for my delicious cowburger to be cooked, I felt driven to analyze the clock on the wall instead of just staring at it impatiently (I was late as usual). What I discovered was a series of numbers:

6 6 9 3

How? I divided the clock into 3 equal pieces (with my powers of visualization, not a hacksaw - no clocks were injured in the making of this theory).

I cut a clock pie into 3 pieces with an arrow at the end of each cut, pointing at the 12, 4 and 8. Adding those three numbers together, you get 24. Adding together the 2 and the 4 from the 24, you get 6.

The next step would be to repeat the procedure with the arrows all rotated by one digit: 1 plus 5 plus 9 equals 15; 1 plus 5 equals 6. That makes two sixes in a row.

Repeated at the next rotation, 2 plus 6 plus 10 equals 18, and 1 plus 8 equals 9.

Again at the next rotation: 3 plus 7 plus 11 equals 21, and 2 plus 1 equals 3. This is the last step, because the next step would be 4 8 12 but that's the same as the first step.

The final result is a repeating sequence of 6 6 9 3:

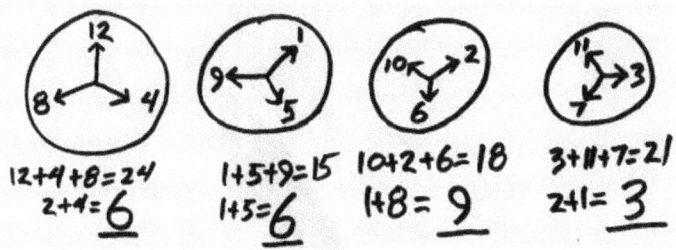

There sure is a suspicious abundance of 3, 6 and 9 going on around here!*

---

* See Appendix Φ, "Pythagorean Skein Reduction Of Successive Decimal Depths of the Golden Ratio."

Appendix 6693

Ancient Solfeggio Allows Time To Exist?

The ancient Solfeggio matrix has some hidden numerical patterns that are only revealed through long division. In particular, Fa divided by Mi reveals a repeating pattern of 6 6 9 3.* This is the exact same sequence as revealed in the aforementioned *clock pie!*

Whoah, dude! Another key to the universe has been found.

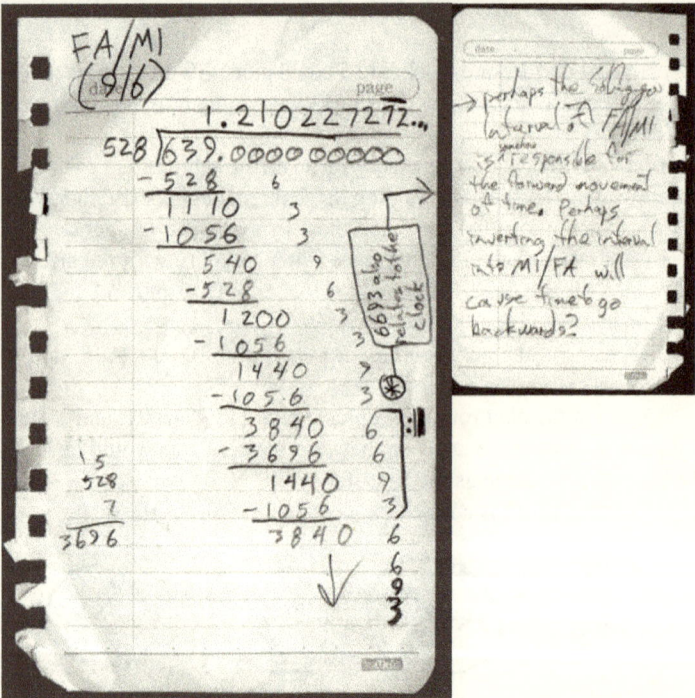

Perhaps the Solfeggio interval of Fa:Mi is somehow responsible for the forward movement of time. Perhaps inverting the interval into Mi:Fa will cause time to flow backwards and enact universal reversal of all entropy...

---

* Matter-o-fact, *all* Solfeggio long-divisions stream a viny waterfall of 3, 6 & 9 below the division bracket!

Appendix Φ

Pythagorean Skein Reduction Of Successive Decimal Depths Of The Golden Ratio

1.6	PSR = 7
1.61	PSR = 8
1.618	PSR = 7
1.6180	PSR = 7
1.61803	PSR = 1
1.618033	PSR = 4
1.6180339	PSR = 4
1.61803398	PSR = 3
1.618033988	PSR = 2
1.6180339887	PSR = 9
1.61803398874	PSR = 4
1.618033988749	PSR = 4
1.6180339887498	PSR = 3
1.61803398874989	PSR = 3
1.618033988749894	PSR = 7
1.6180339887498948	PSR = 6
1.61803398874989484	PSR = 1
1.618033988749894848	PSR = 9
1.6180339887498948482	PSR = 2
1.61803398874989484820	PSR = 2
1.618033988749894848204	PSR = 6
1.6180339887498948482045	PSR = 2
1.61803398874989484820458	PSR = 1

...ad infinitum

A number series is created:

    7 8 7 7 1 4 4 3 2, etcetetra...

I don't see any repeating pattern. So, what does it mean?

(continued on next page)

Perhaps reducing the skein series a little further will reveal some sort of universal mystery:

```
7 8 7 7 1 4 4 3 2 9 4 4 3 3 7 6 1 9 2 2 6 2 1 ... ad infinitum
 6 6 5 8 5 8 7 5 2 4 8 7 6 1 4 7 1 2 4 8 8 3 ...
  3 2 4 4 4 6 3 7 6 3 6 4 7 5 2 8 3 6 3 7 2 ...
   5 6 8 8 3 9 1 4 9 9 9 3 2 3 7 1 2 9 9 1 9 ...
    2 5 7 2 3 1 5 4 9 3 5 5 1 8 3 2 9 1 1 ...
     7 3 9 5 4 6 9 4 3 8 1 6 9 2 5 2 1 2 ...
      1 3 5 9 1 6 4 7 2 9 7 6 2 7 7 3 3 ...
       4 8 5 1 7 1 2 9 2 7 4 8 9 5 1 6 ...
        3 4 6 8 8 3 2 2 9 2 3 8 5 6 7 ...
         7 1 5 7 2 5 4 2 2 5 2 4 2 4 ...
          8 6 3 9 7 9 6 4 7 7 6 6 6 ...
           5 9 3 7 7 6 1 2 5 4 3 3 ...
            5 3 1 5 4 7 3 7 9 7 6 ...
             8 4 6 9 2 1 1 7 7 4 ...
              3 1 6 2 3 2 8 5 2 ...
               4 7 8 5 5 1 4 7 ...
                2 6 4 1 6 5 2 ...
                 8 1 5 7 2 7 ...
                  9 6 3 9 9 ...
                   6 9 3 9 ...
                    6 3 3 ...
                     9 6 ...
                      6 ...
```

Well, if nothing else, there sure does seem to be a lot of 6, 3 and 9 at the bottom of the funnel. Hmmm...... 3, 6, and 9, huh? *Very* interesting! 3, 6 and 9 seem to pop up a lot during the adventures in this book* - their ratios describe the octave and the perfect fifth (3:6:9 = root:octave:fifth), and are an important part of Solfeggio theory, among other things.

---

* See also Appendix 6693, as well as Appendix 6693.

Appendix 1497

Palindromic Skein Sequence In The Squares

Discovery par moi : the skein reductions of the squares...

x	x^2	Skein
1	1	1
2	4	4
3	9	9
4	16	7
5	25	7
6	36	9
7	49	4
8	64	1
9	81	9
10	100	1
11	121	4
12	144	9
13	169	7
14	196	7
15	225	9
16	256	4
17	289	1
18	324	9
		etcetera...

....produce a palindromic number series (1 4 9 7 7 9 4 1), repeatedly bookended by mirrors of 9...*

1 4 9 7 7 9 4 1 **9** 1 4 9 7 7 9 4 1 **9** 1 4 9 7 7 9 4 1 **9** etcetera...

...!!!!!!!!!!!!!! WOW !!!!!!!!!!!!!!...

Considering the missing 9 at the beginning of the sequence, I suppose it goes to follow that the skein reduction of 0^2 should be...

9

...????????????? WTF ??????????????...

---

* In the realm of enzyme structures: of all the possible 2-way combinations of the letter bases *A, C, G* and *T*, the duplet *AA* has a special function as *punctuation mark*, signaling the end of one code and its separation from the next code - much like 9's function here.

Appendix 364.5

The Orbital Gnomon Star Power Chord

And now I present another big discovery of mine, involving angles this time. More late-breaking, ground-breaking news from the land of the Golden Number. I've found a three-way link between Earth's orbital period, phi, and the musical perfect fifth.

Okay, so...

Picture if you will, a star...

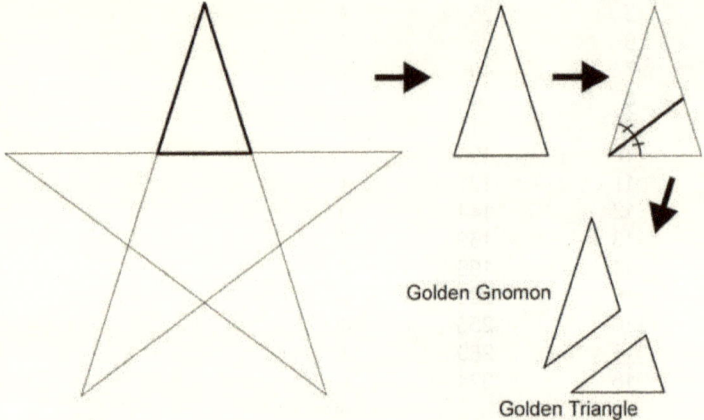

Start with the topmost triangle of a perfect pentagram. Bisect one of its base angles. This splits the triangle into two pieces, which happen to be called the *golden triangle* and the *golden gnomon*. These names are irrelevant to the process, just trivia for trivia-curious readers or authours.

What I am about to reveal to you next is my very own discovery, and it is an extension of the star-splitting procedure pictured above.

In my hopeful search for some sort of treasure of interconnection, musical or otherwise, I whipped out the old calculator watch (it really is getting old - the numbers have worn off and the strap is now broken) and divided a couple of angles here and there...

SECTION VI: APPENDICES

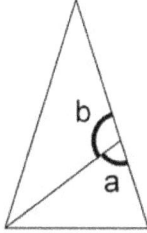

b / a = Perfect 5th

With joy, my wish was swiftly granted. When I discovered that angle "b" divided by angle "a" equals the musical perfect fifth ratio of 3-over-2, I almost crapped my pants! But not in the same way as with a bad hot dog - it's just a figure of speech.

I took this as a sign and thought I should re-iterate the transformation, in the same way that the Circle of Fifths is generated by the reiteration of the perfect fifth in the musical realm.* It just seemed like the right thing to do; starting with angle "a," which is 72 degrees, I applied the musical perfect fifth repeatedly by multiplying by 3/2:

$$72 \times 3/2 = 108$$
$$108 \times 3/2 = 162$$
$$162 \times 3/2 = 243$$
$$243 \times 3/2 = 364.5$$

I was pleasantly shocked to see the number 364.5 show up suddenly like that... *very* close (within 0.2% accurate) to the orbital period of planet Earth, which is 365.25 days.

Therein lies the incredibleness. It seems to me that *planet Earth's year is harmonically tuned to the Golden Ratio by a series of musical perfect fifths.* The infrastructure of our Universe seems to be built upon cross-disciplinary metaphors. Here we have three elements mysteriously connected.

planetary motion : musical interval : golden ratio

I name this phenomenon the *Orbital Gnomon Star Power Chord*.

May this new discovery of wisdom forever shine its radiant light on cosmic existence.

---

* Refer back to p. 39 for the details.

Appendix 365

Gas Bill

The gas bill pictured below, which I received in my mailbox just today, has an amount due which is *very* close (within 0.058%) to the orbital period of planet Earth, which is 365.25 days.

Billing Date: Aug 10, 2004		
Account Number	**Due Date**	**Amount Due**
	Sep 1, 2004	$365.46

Are some coincidences more meaningless than others, or am I just passing gas here?

...and don't even get me started on this here ATM receipt:

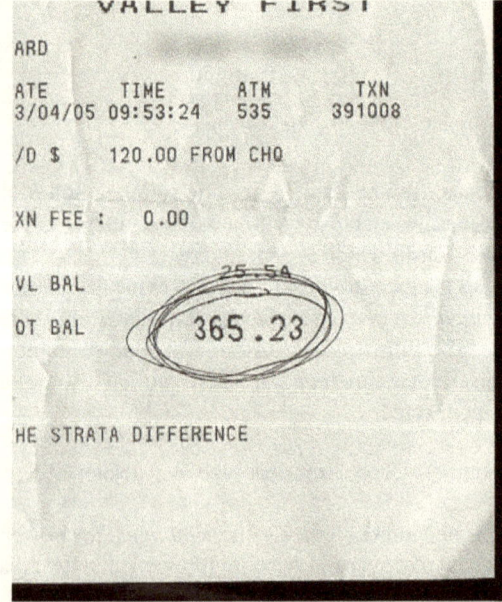

...99.994% synchronistic to the Earth's orbital period! I am obviously a cosmic wizard of serendipity when it comes to things like financial statements.

Appendix ±

The Margarine Era

As if the question of whether numbers exist in the first place isn't trivializing enough already, what about all the errors involved? In real natural phenomenon, hardly any of these numerical whatchamacallits ever appear in perfect form. There always seems to be some inaccuracy involved; it's almost as if all the fancy formulae are just "tendencies" rather than "rules." Something always falls out of kilter; this slight lack of focus is commonly called "margin of error" or, as I like to call it, "The Margarine Era."

You could call it the sloppy joe factor. In some circles, the margin of error is called *tolerance*, and indicated by the sign "±" followed by a dimensional quantity which tells how much error is allowable in the design - isn't that great? Our world could use a little more tolerance.* Maybe then we wouldn't have so much pain & war & death & famine & crap like that.

Example: "I woke up at noon today (± 1 hour)" means that I didn't exactly obsess over the specifics of the temporal location in question but the picture is somehow assumed to be accurate enough for the given situation. Special comittees are responsible for these sorts of specifications whenever necessary.

In astrology, there is a term for the margin of error when assessing the alignments between various celestial bodies in a chart - this range of tolerance is called the *orb* and is usually 8 degrees or less. As a percentage expression, I suppose this would be eight divided by one-hundred-and-eighty, since the furthest away you can get on a circle is halfway around. That's a tolerance of about 4.4444444444%.**

In the math of music, margin of error occurs naturally in the area of tuning. If I had a guitar pick for every time I've heard someone say, "What do you mean my guitar's out of tune? Close enough for rock and roll, man!" I could open a music store specializing in plectra. But, *really,* even a simple chord played on a guitar with perfect intonation that is tuned to an obsessively meticulous level of accuracy and played at a consistent loudness by a knowledgable and experienced so-called expert in a steady-temperatured environment would still inevitably have a chaotic string of irrational non-zeroes on

---

* November 16 is the official International Day Of Tolerance. See you there?

** See Appendix 4 as well as Appendix 4:44:44 04/04/04 if you enjoy stories about the number 4.

the right side of its tone-division decimal point. So, again, perfection seems to be this elusive, impossible ideal. Is this really such a bad thing?

In fact, some tuning anomalies here and there can really make the music sound more *alive* - as if it is *breathing*. The sound can become more resonant, exciting - hypnotic, even. If you listen to the sitar, one of the most beautiful noisemakers on Earth, it sounds great and creates good feelings amongst most who hear it. And, although I am no sitar expert myself (yet), (or is it: ",)" ?), I can guarantee that the tuning of all its various strings are not tuned to 100.00000000000000000% accuracy to the exact pitch in cycles-per-second as specified in the policy handbook. If it were, it wouldn't purrrrr.

Ergo, the Margarine Era can be a big bonus. Another great analogy comes from the graphic design world. It's a fancy little phase-cancellation thing called *moire*. Moire* is the pattern that magickally appears from out of nowhere when several grid-formations of dots are superimposed, with the grids at various angles:

 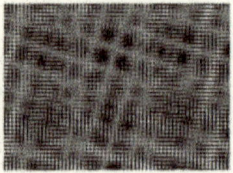

The picture on the left has the dots all at the same angle, resulting in a dumb, boring old grid pattern. The picture on the right is moire. We want the moire! Down with perfection! Then again... I suppose plaid can be sexy in its own unique way, too.

In relation to the Solfeggio tones, the Margarine Era spells nothing but havoc for the underlying mathematical system that relies so heavily on the numerological reduction of its frequencies. We ought to be able to achieve perfect Solfeggio music without computers, but *how*?

I mean, who could sing or chant an exact three-hundred-and-ninety-six-point-zero-zero-zero-zero-zero-zero-zero-zero-zero cycles-per-second?

---

* *Moire* originates from the French *moiré*, meaning "having the appearance of watered silk." In audio engineering there's a phenomenon called phase-cancellation, which when embodied via the psychoacoustical *Haas effect* or other forms such as *chorus* or *flange*, creates interesting 3-D sorts of effects across a mere stereo field for the listening audience. These applications of phase modulation are somewhat equivalent to putting a reverb, and** a *pre*verb, to the phases of the Moon!

** ...subsequently and cyclically, with an optional bit of "simultaneously" thrown in for more of a spinning-frilly-skirt sort of swish!

Appendix $Z = Z^2 + C$

Mandelbrot Doppelgänger

At this one place I used to live, there was a paintchipped dent on the front of the oven door. The damage was there before I arrived. The real reason I took a photo (regardless of damage-deposit documentation) was 'cause it's a dead ringer for that classic example of fractal generation, the good ol' Mandelbrot set:

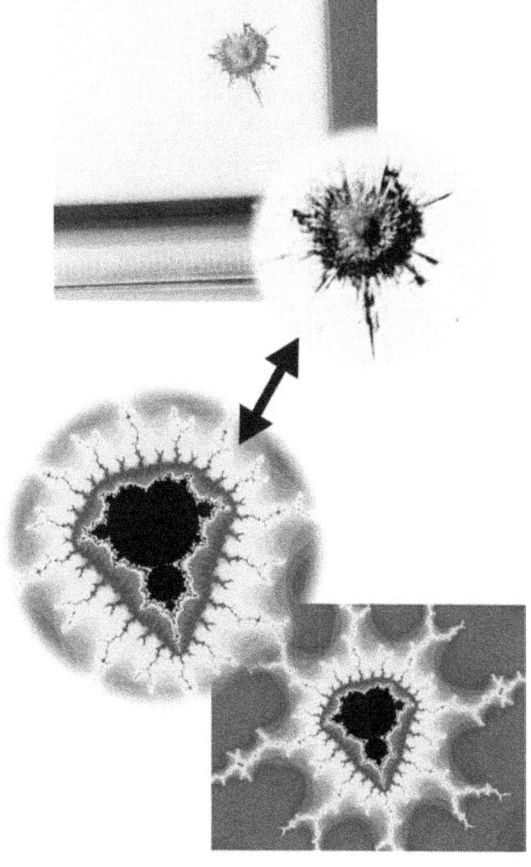

A near-perfect replica. Pretty neat, huh?

Appendix 6.4

Koch Snowflake

The following procedure illustrates the jist of *fractals*.

1) Start with a big old equilateral triangle:

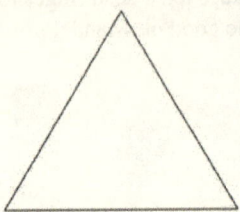

2) Add a new, smaller equilateral triangle to each side of the first triangle:

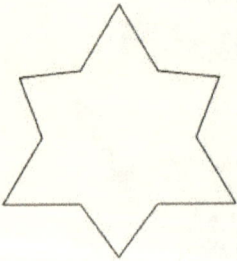

3) Continue the process by adding more new triangles to the middle of each side of every new triangle:

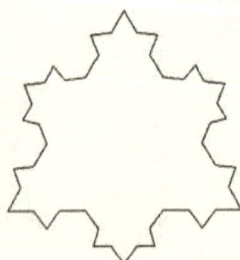

4) Repeat ad infinitum. An infinitely complex and finely detailed shape known as the Koch Snowflake is created. No matter how far you zoom into the Koch Snowflake with a microscope, there will still be finer levels of embellishment that defy perception except at the level of the unattainably infinitessimal:

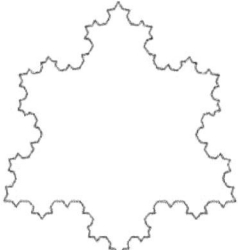

This shape has some interesting properties:

a) Its perimeter (length of string along the entire curve) is *infinite*, whereas

b) its area approaches (but never reaches) a *finite* limit which is equal to 8/5 of the area of the original triangle. Eight-fifths? Eight and 5 are both Fibonacci numbers! Wow! Also,

c) with every step of the procedure, the perimeter increases by a factor of exactly 4/3: a musical *perfect fourth!*

Double wow!

As if all this musicality isn't incredible enough already, the Koch Snowflake is extra-great because its model of creation symbolizes fractals in general.* Fractals are mathematical number systems which are self-similar and scale-independent...** and they describe a lot of natural phenomena such as tree roots and branches, cloud formations, turbulence, the Stock Market, coastlines, tide patterns, and more!

Nature wants to be understood, otherwise everything wouldn't be so simple.

---

* The Mandelbrot set, showcased a couple of pages ago, is probably history's most famous and recognizable example of a fractally-generated graphic, having showed up on countless posters, album covers and t-shirts over the last couple of decades.
** It is even possible for *sound* to display the property of scale-independent self-similarity. Mandelbrot called them scaling noises: *white noise, Brownian noise,* and *1/f noise* are examples.

Appendix 6.5

Koch Tetrahedron

As previously described, the 2-dimensional Koch snowflake is a great example of a fractal doing what fractals do best. Which is, of course, to unfold deeper and deeper layers of the fine fabric of reality, zoning in closer and closer to the all-pervading yet unquantifiable source. One day I found a two-dimensional fold-up pattern for a tetrahedron - in a book about *words*, of all places. As yet another great example of "as above, so below," the cut-'em-and-fold-'em pattern for a tetrahedron has the silhouette of an equilateral triangle:*

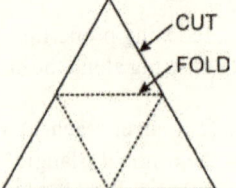

I spontaneously put this together with the Koch snowflake concept, and visualized a tetrahedron giving rise to smaller and smaller tetrahedrons at the center of each face of the "root" tetrahedron - repeatedly adding more and more smaller "overtone" tetrahedrons, growing like a harmonic crystal.

Of course, I had to write down my amazing new discovery.

"I could become famous for this one," I assured myself. I even drew a picture of what the first few iterations would look like:

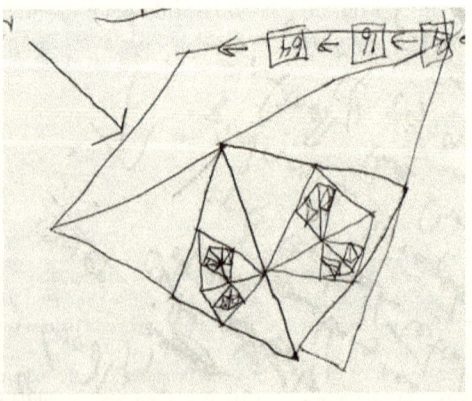

---

* The diagram on p. 93 *begs* to play in this game, does it not?

The timing of this discovery was serendipitous - it was a full moon, payday, Cinco de Mayo, Beltane, a friend's birthday, the 100th birthday of my piano, and Venus was shining particularly bright on that night.

The next day, after a quick search on the 'net, I found that my new theory was old news. Math students have probably been learning this stuff for thousands of years already. Dammit! I thought I was onto something big here, like I was gonna get all this recognition for expanding the collective mind with my amazing discovery, not to mention the monetary reward, which is, well, more than you can possibly imagine.

And I can imagine a lot.

Oh, well (patting my own back like an Oroborus)... it was still cool to have that eureka moment and connect that new synapse all on my own.

After recovering from the disappointment of my non-originality, I continued to read through the reference material and I learned something really wild: the eventual shape of the Koch tetrahedron process is...

Are you ready?

Take a guess... If you started with one large tetrahedron and added a new tetrahedron to each side, and added more new tetrahedrons to every face of the previous generation of tetrahedrons, forever and ever, what would the ultimate shape turn out to be? Would the tetrahedron just keep getting bigger and bigger, or would it turn into a different shape altogether?

The ultimate fruition of the Koch tetrahedron is... a CUBE!

How's *that* for evolution? Isn't that *amazing?!*

I probably woulda never guessed that one in a million years...

Awesome.

Appendix 6

Shine On, You Crazy Tetrahedron

Apparently, diamonds have a molecular structure which echoes the shape of a tetrahedron, with a Carbon atom at each corner and one at the center!

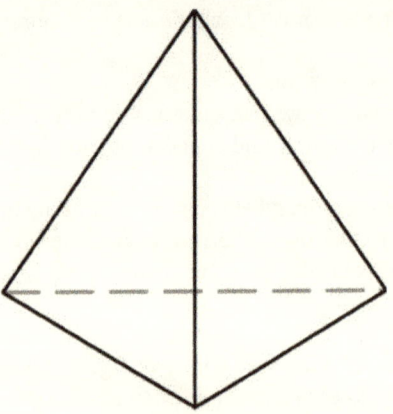

Tetrahedron

Just last night I heard a great story about a new discovery via the Hubble Telescope. Apparently, there's a burnt-out star somewhere out there that used its own gravity to crush itself into a huge diamond, 2.5 miles long.* That's a heck of a lot of tetrahedrons.

Also, diamonds are being continually produced at the centres of the planets Uranus and Neptune. This assertion is based upon experiments conducted here on Earth where liquid methane is put under huge amounts of pressure.

Liquid methane?

Wasn't there a thing in the news a few years ago about how too many cows down at old MacDonald's rainforest burger farm are farting too much methane into the atmosphere and wrecking the ozone layer? *Hmmmmm...* (dollar signs in my eyes) maybe we could train cows to squeeze their butt-cheeks together really hard and produce diamonds for us...

"'Scuse me, Miss Jersey, how are you today? Could I please pull a diamond out of Uranus?"

---

* ±.

Uranus - the black sheep planet nobody wants to talk about for fear of indecent innuendo. Is it urinating on us or is it checking out your anus? Oh, the shame! Hey Uranus,* I heard Pluto,** Re*** and Indigo**** are starting a *freaky weirdo club*, why don't you go join it?

[They can make me grow old but they can't make me grow up.]

Oh, and apparently, in February of 2004, astronomers announced that a burnt-out white dwarf named BPM 37093 is actually a 10-billion-trillion-trillion carat diamond.

Excuse my French, but that's beaucoup des zeroes, and a lot of frickin' tetrahedrons, too.

Apparently, burnt-out white dwarves are a girl's best friend.***** Okay maybe the puns are getting out of hand here.

Unhand that pun!

Appendix 8

Allotropic Faces

In chemistry, the two *allotropes* of carbon are *diamond* and *graphite* - two physically different manifestations of a single chemical entity. Diamond and graphite are like day and night, Jeckyll and Hyde. Diamond is clear and hard and useful for impressing people who like tetrahedrons or shiny things; graphite is flaky and black and useful as lubricant (for machines, not people).

I see a great metaphorical connection between the allotropes of carbon and the oscillations of 6 and 3 in the Solfeggio Transposition Triangles. Carbon has two faces, and so do Re, Mi, Sol and La. When Mi's skein is 6, for example, we could say it is in its diamond form. When it is transposed any odd number of octaves up or down and its skein becomes 3, Mi wears its graphite hat. Therefore Re, Mi, Sol and La each have two allotropes: diamond and graphite.

---

* Did you know that Uranus was actually named "George" by the astronomer who discovered it? Did you also know that Englebert Humperdinck is not Englebert Humperdinck's original name?
** See p. 19.
*** See p. 102 and p. 105.
**** See p. 71.
***** Hmmm... and dogs are sometimes called "man's best friend" - dogs, dwarves, diamonds... there's gotta be a song in there somewhere!

Appendix 7

The Two-Tone Tuning Fork

In all my years of playing guitar and other musical instruments, I have never had my ears so obviously tricked as just a couple of weeks ago. Fooled by a tuning fork, of all things.

On the edge of a table I frappe a tuning fork, and with my other hand I strike the bars of a trusty xylophone to compare the pitches. With superhuman ease and godlike musical stealth I determine the tuning fork to be perfectly tuned to the C note of the xylophone. I tap it again, looking forward to a close-up listen. As I bring the tonal cutlery closer to my pinna, a miraculous transformation occurs - the C changes to E! I have never ever witnessed such a tuning fork as this, and I have been playing guitar *forever* (almost 19 years). As it turns out, the tuning fork was really an E-fork, not a C-fork. In a state of disbelief, confusion and awe, I tap the tangs on the table again.

A two-tone tuner. Wow! C above E: a minor sixth. I'm totally shocked that:

a) this tuning fork sings two different notes, depending on its positioning in relation to the ear, and

b) the interval of minor sixth resonated so prominently, rather than a lower-numbered ratio such as a perfect fifth or fourth.

Shocked, but certainly not disappointed. Bewilderment: a privilege or a right? Discuss amongst your elves.*

Appendix 1:2:3

Mercury Tetractys

Planet Mercury perfectly embodies the overtone series, whipping around the Sun singing harmonies with itself all day long. Mercury's day is twice as long as its year, creating a perfect octave. Also, Mercury rotates on its own axis one-and-a-half times per Mercury year: the musical perfect fifth!** The cosmic architect is *definitely* a musician.

---

* Refer to "twelfth" in the Glossando for pertinent discussion.
** Mercury's year equals 88 Earth days - the same number of keys on a standard piano! Mythological Mercury, with his caduceus wand, was hailed as a god of music.

Appendix 11:22:33:44:99

The Sunspot Dozen

11: The Sun has an 11-year sunspot pulse.* Jupiter - whose orbit is also 11 years - is 11 times the size of Earth.**

22: The number 22 is an octave of 11; the Hebrew alphabet has 22 characters, and the Internet told me that Indian classical music once upon a time divided the octave into 22 steps.

33: The intervals of the ancient Solfeggio matrix feature the number 33 in a couple of instances; 33 is a musical perfect fifth compared to 11 or 22. Also, according to the book *11:11,* 33 is the "master vibration number of Universal Service."

44: The ancient Solfeggio also contains the number 44 in its golden interval of Fa-over-Ut; 44 is yet another octave of 11.

99: Ninety-nine bottles of beer on the wall.

Who knew 11 was so rockin'-ly awesome? Maybe we need a new dozen! We've already got the standard dozen of 12 and the baker's dozen of 13... an 11-based dozen just might serve some demand.***

---

* The Sun (a.k.a. *Sol*) also has a 9-times-a-day sonic cycle. That's 160 minutes per cycle - the same number of rooms in the Winchester Mystery House! Cosmic coincidences never cease. If the 160-minute cycle is transposed up by exact *octave* increments (22 of them), its corresponding musical tone (436.9 Hz) is found to be very close to the A note (436.05 Hz) of the Just Temperament system. According to my explorations in Section IV, this frequency corresponds to the colour *orange*.

** The Sun and Jupiter could be considered a double Sun, just as Earth and Moon are a double system and so are Pluto and Charon. And so I ponder... does Jupiter stir up the sunspots into existence?

*** Further cosmic support for 11... the diameters of Earth and the Moon are in 11:3 ratio; also, if you divide the furthest distance between Venus and Mars by their closest distance, you get the same 11:3. And so I ponder... since all ratios in the Universe beg to be compared to the ratios of Just Temperament intervals, what is 11:3's equivalent in the language of musical interval? The answer: halve the fraction to limit the result between 1.0 and 2.0 as per the previous discussions in Section III, so 11:6 becomes the real question. Eleven divided by six equals 1.8333... This number is partway between the minor seventh (1.800) and the major seventh (1.875), slightly closer to the minor. Who knew?

Appendix 144

12-Bar Blues

Once upon a time, a dismembered hotel Bible brought me an omen involving the number 144... something to do with the dimension of heaven, if I understood correctly... And so I pondered, *"144...?* That's a perfect square - twelve times twelve - just like the 12-bar blues in 12/8 time!"

Please let me explain. One of the bestest, most popular song forms in the history of planet Earth is the *12-Bar Blues*. I invented it myself - the same day I invented heavy metal, actually - and its basic symetry looks like this:

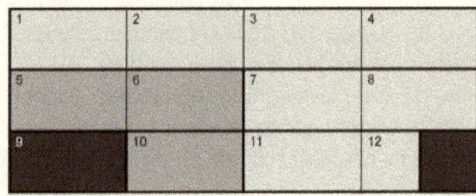

Legend:

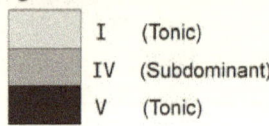

I (Tonic)
IV (Subdominant)
V (Tonic)

Another common musical occurence of the number *12* is the 12/8 time signature, also known as 4/4 "swing" time. As the numbers imply, 12/8 time has twelve eighth-notes per measure - 12 small time-units per larger time-unit; twelve spokes of a wheel. This is a very popular form of rhythm. Without the "swing" parameter, the wheel of 4/4 time is divided into 1, 2, 4, 8, 16, 32, 64, 128, 256, or 512 parts - depending on how far you zoom into the fine fabric of rhythmic reality. This is not the same ball of wax as the 1, 3, 6, 12, 24, 48, 96, 192, 384 pieces of the 12/8 pie. Other time signatures do exist; listen to Rush, Frank Zappa, jazz and prog-rock for examples.

A favourite activity of mine is to play 12-Bar Blues in 12/8 time, thereby combining the two different 12-isms into a new matrix:

	1	2	3	4	5	6	7	8	9	10	11	12
Bar 1	1	2	3	4	5	6	7	8	9	10	11	12
Bar 2	13	14	15	16	17	18	19	20	21	22	23	24
Bar 3	25	26	27	28	29	30	31	32	33	34	35	36
Bar 4	37	38	39	40	41	42	43	44	45	46	47	48
Bar 5	49	50	51	52	53	54	55	56	57	58	59	60
Bar 6	61	62	63	64	65	66	67	68	69	70	71	72
Bar 7	73	74	75	76	77	78	79	80	81	82	83	84
Bar 8	85	86	87	88	89	90	91	92	93	94	95	96
Bar 9	97	98	99	100	101	102	103	104	105	106	107	108
Bar 10	109	110	111	112	113	114	115	116	117	118	119	120
Bar 11	121	122	123	124	125	126	127	128	129	130	131	132
Bar 12	133	134	135	136	137	138	139	140	141	142	143	144

This matrix is 12-by-12 square, with a total of 144 beats in the pattern. Measured against the Earth's energetic gridwork of leylines, the speed of light is 144,000 miles per "grid-second." In classic gematria (Hebrew numerology) the number 144 represents *light,* as well as the alignment of the micro- and macro- (as above, so below). The Fibonacci series contains the number 144. There are 1440 minutes in a day. Also, the Mayans divided each day into 144 "Mayan minutes."

I propose an extrapolated version of the standard chess game which would use a 12x12 board instead of 8x8; four extra pawns on the frontline plus two instances of two newly-invented special-ability characters on the backline oughta do the trick.

Lots of "144" stuff to get excited about... awesome, eh?

Appendix 420

Re-Factoring The Calendar

One fine day, inspired by the notion that Chinese mathematicians divide the circle into 365 degrees to match the number of days in a year, I performed yet another coffeeshop-type research adventure with the intent of collecting some possibilities of circle divisions. Many of us are familiar with 360 degrees per circle... I wanted to derive alternative division systems for the circle. At first, I was intent on finding a division of 13, to reflect the fact that the Moon orbits Earth 13 times per year.

I came up with 390, 351 and 273 degrees per circle. 273 I found awkward to relate to and remember for some reason, yet interesting in its construction because it separates into 13-times-21 (and I like 21 because it's 3 times 7). 351 is also kinda awkward but interesting because it separates into 3 times 3 times 3 times 13. 390 seemed like a much smoother activation of the 13 vibe, both easy to relate to and divisible by convenient and popular numbers such as 2, 3 and 5.

936 is 2x3x12x13, bringing the 12 and 13 together in a destined and feasibly important harmony. It also belongs to the elite and mysterious club of numbers that include the digits 3, 6, and 9.

Since the diameters of Earth and Moon are in a 11:3 ratio, I attempted to apply that ratio to some other sensibly-factorable number, such as that old kleeshay, 90 degrees, to arrive at a new division of the circle. 11:3 times 90 makes 330 degrees per circle. I noticed that 330 factors out to a similar-looking set of numbers (2x3x5x11) as 390 (which is 2x3x5x13), thereby perfecting their function as parallel representations of the resonances 11 and 13. Maybe we need an 11- or 13-hour clock as well as our 12-hour standard?*

The answer to the question, "how much does the Earth's diameter increase when the Moon's diameter is added to it?" is 14-over-11; so I thought it would be cool to multiply this number by the 330 which I found earlier. 14:11 times 330 equals 420. Since the 11 that appears as a factor of 330 is cancelled out through multiplication by 14/11,

---

* The ancient Egyptians had three separate calendars, each with practical relevance and unique cosmic basis - one based on Moon cycles, one based on Earth's 365$^{1/4}$-day cycle around the Sun, and a 365-day cycle called the "civic" calendar. The civic calendar lines up with the Sun-based "Sothic" year every 1461 civic years at a convergence which they called the "New Year," and twenty-five of which correspond to exacty 309 full-moons.

the resonance of 11 seems to function as an idler-gear in the transformation. 420 separates nicely into a great many factors: 1, 2, 3, 4, 5, 6, 7, 10, 12, 14, 15, 20, 21, 28, 30, 35, 42, 60, 70, 84, 105, 140, 210. This is the same level of factorability as the current 360-degree standard; therefore, 420 degrees-per-circle should probably be considered as a viable alternative for the compass and the clock.

Geometry would gain popularity quite quickly, I'm sure, if there were 420 degrees in a circle.

And, another question: is high factorability - such as is offered by the 360- and 420-degree circles - really all it's cracked up to be? Maybe the convenience of it all reeks of a sourly-tuned cop-out - not unlike the Equal Temperament system.

Appendix 13

The Fat Orphan

I overheard a terrible conversation at the pool* the other night. It was so perfect that it must have been planned by spies, like they were saving the story just for "the weird guy who's into the numbers and stuff." These two girls were sitting next to me in the sauna, and one tells the story of a friend of hers, who is one of twelve blood-siblings. The family adopted a thirteenth child. The adopted child happened to be a fat kid, and so the other sisters and brothers made a habit of calling him "mad cow disease."

Some people are so mean. I should have said something like, "that's mean" or "the 13th bun in a baker's dozen has special significance, you know. It's a safety factor, worthy of respect. Don't you shitheads know the Moon orbits the Earth 13 times a year?** Maybe your friends should be nicer to their new brother," but instead I just sat there in the boiling heat and sweated out the triskaidekaphobic ugliness of humanity, not unlike poisons leeching from a gangrening wound.

---

* See Appendix 0.55555 for another numerical synchrodipity story involving the public pool.

** Did you know that Sarah Winchester had the 12-candle chandelier at the Winchester Mystery House changed to 13? Also, the coat-hooks are in multiples of 13, and the fancy Tiffany window has 13 stones.

Appendix 33

Solfeggio Factors

Ut 396
198 x 2
132 x 3
99 x 4
66 x 6
44 x 9
36 x 11
33 x 12
22 x 18

Re 417
139 x 3

Mi 528
264 x 2
176 x 3
132 x 4
88 x 6
66 x 8
48 x 11
44 x 12
33 x 16
24 x 22

Fa 639
213 x 3
71 x 9

Sol 741
247 x 3
57 x 13
39 x 19

La 852
426 x 2
284 x 3
213 x 4
142 x 6
71 x 12

Appendix S

Entropy

Please enjoy the following illustration of increasing entropy:

Increasing Entropy

Kudos to the music cassette tape for contributing its organs to science.

Entropy is a measure of "how much disorder" in any given situation. Examples of "increasing entropy" include a hot stove element cooling down, a rock falling to the ground, a drop of colour dissolving into a cup of water, and sandcastles washing away in the tide. Examples of the "decreasing entropy" concept include building something, making time move backwards,* freezing water, and squeezing a lump of coal into a tetrahedron with your bare hands.

However, even when attempting to create order, extra energy is always used, directly or indirectly, which actually results in the overall entropy level of a situation being higher than one might have expected. Science calls this the "second law of thermodynamics," which says that "entropy always increases in a spontaneous system," or "energy spontaneously disperses from being localized to becoming spread out if it is not hindered."

For example, if I wanted to attempt a decrease in entropy by re-assembling the magnetical dog's breakfast shown on the previous page into its useable original form, I would either be spending a lot of time and aggravation, not to mention the food intake to provide the energy of doing the work, or else spending even more resources to build a machine to do it for me. How many crumpled papers and discarded prototypes worth of no-longer-useable material will it take to finally perfect the machine, and how much hazardous waste will get spilt into the ecosystem by the factory which helps builds the parts which I designed to produce my tape-winding machine which I will then use to minimize the entropy I create during the reconstruction of the cassette tape?

What a paradox! The harder we try to create order, the quicker it crumbles like a bad joke falling under the irresistible pull of a black hole. If technology really is going to save us, it certainly has its price.

Are the advantages of technology worth its inherent entropic madness?

---

* See Appendix 6693.

Appendix S-VHS

Entropy (VHS Remix)

...'nuff said.

Section VII

GLOSSANDO

Vocab words.

abacus

-noun Beads sliding across wires to represent numbers and perform calculations thereof... just imagine how much more in tune we all would be with the world around us if we moved the numbers around for real instead of getting the electrical calculator to do all the thinking on our behalf! Etymology inquiries point to *sand*, and it only makes sense (to people familiar with sand, anyways) that drawing lines in the sand with one's finger is the most sensible, self-sufficient, ergonomic, and environmentally responsible method of performing calculations. Era- and culture-specific abacus-ological variations abound regarding various details such as the number of beads above and below the center beam, the existence whatsoever of said such beam, and the x-versus-y directionality of the sliders. For example, the modern Japanese *soroban* uses four beads below the beam and one bead above - what a great parallel to the humand hand!

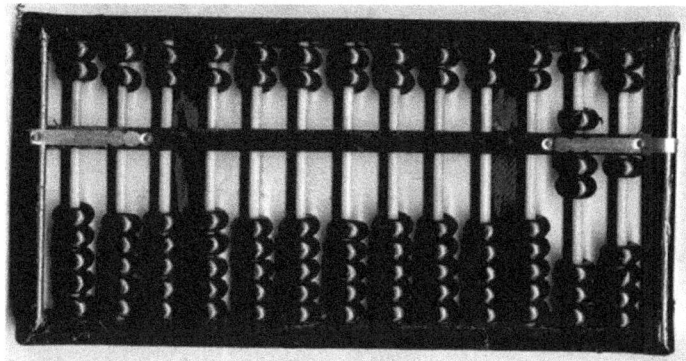

FIG. 4: A REASONABLE FACSIMILE OF MY BRAND-NEW ABACUS - I JUST BOUGHT IT TODAY. CAN YOU TELL WHAT NUMBER IS BEING SHOWN?

axiom

See *maxim*.

paradox

-noun A contradiction or seeming impossibility. *(eg. The present tense is simultaneously inescapable and always escaping. The "present" keeps escaping. You can't hold on to it because it keeps slipping through your fingers. However, the present is the only thing that actually exists, so, like, what the... ?)* See also *pair of ducks*.

facetiousness
-*noun* The quality of having a thinly veiled second meaning; sometimes similar to sarcasm but not quite as bitter. Calling the kettle black is sometimes an act of facetiousness rather than one of ignorant hypocrisy, and therefore becomes somewhat comedic, like a moebius oroborus mime standing in front of a mirror. *(eg. "If you want to be eloquent like me then don't f#*kin' swear so much!")*

moot
-*adv.* Having far less practical than hypothetical relevance; most often used to describe an argument or issue.
For example, asking where to bury the survivors of a plane crash might be considered a moot question.

the other hand
-*noun* On one hand, the pedestrian who looks over his shoulder all the time may appear paranoid; on the other hand, the pedestrian who *doesn't* look over her shoulder all the time might get squished by any of the umpteen maniac drivers in this town. What we need is a public racetrack, with a waiver of course, where we can all go apeshit with our nitrous oxide turboes, etcetera.

orbital radius
-*noun* The average distance of a planet from the Sun on its yearly travel.

dervishes
-*noun* I'm probably not the only one who ever got drunk and spun around twenty times while hunched over and running with a baseball bat hanging from my forehead, then trying to run across the lawn without skidding out and puking all over the place. It's its own special kind of "talking to God" kind of game. This may or may not be similar to the experience to be had by whirling like a Whirling Dervish, but it looks like fun anyway. Except - how do they not fall down or bump into each other? I guess the cone-shaped pants are like bumper-guards, or warning devices.

mosh
-*noun* A mosh pit.
-*verb* To engage in the act of slam dancing. *(eg. "I was caught*

in a mosh" or *"The Slayer pit was fu&%in' wicked, man!")*
The following rules apply to moshing:
0) Most sensible people don't follow rules so maybe just skip this part.
1) If someone falls, help them up instead of stomping on them. Rock and roll is no excuse for abuse.
2) Guns are not invited. No billy clubs or pepper spray either, thanks.
3) Groping is also generally not invited, but then again some would beg to differ. Same goes for item #2 I guess, but then again maybe we should invent a gun-vacuum that sucks all the guns on Earth into a trash compactor and then chucks it into the centre of the Sun for safekeeping.
4) Purposeful non-consensual hurting is not an intended part of the mosh game. However, neither is whining. If you're gonna mosh, don't be too surprised by a bruise or two.
5) No elbowing. And keep your elbows off the table.
6) Stage diving is obviously fun, but it comes with its own set of social and medical dynamics which may be worth considering. For example, if your boots or some other fashion item appears too dangerous, the more fearful of the earthbound moshers might move out of the way and you'll crash to the ground. Try to avoid this. Concrete hurts. Also, security guards are quite often paid to prevent stage diving, presumably at the request of the hall owner's insurance agent. Good luck. In a free world you could do whatever you want - including violate the freedoms of others. *Right? Hey* - wait a minute... is this one of those moebius-22 thingies, or a completely disproportionate and inappropriate political extrapolation? Try to help catch the stage diver if you can. If the load is distributed over several absorbers, the chance of injury is lessened for everyone.

Disclaimer: this literary work and its affiliates do not officially encourage anything dangerous.

disclaimer
-noun Any message which translates into "the buck stops elsewhere."

fig. 4
-abbr. See *figure 4*.

figure 4
-*noun* A brutal wrestling move that could snap one's twiggy chickenlegs like a stale pretzel. If you have nice calves then you may want to insure them before subjecting yourself to this popular technique.

gobbledeygook
-*noun* 1. verbalization of questionably obsessive and perhaps elitist depth 2. jargon 3. fancy talk 4. containing so much extra verbage as to become effectively the same as some sort of alien language that is not understandable to the non-understander - especially when used as a sales tactic. *(eg. The vacuum-cleaner salesman filled my head with so much gobbledygook about suction and pressure and friction and whatever other space-pilot garbage, I had to buy one of the damn machines just to put a stop to all the fancy talk! Maybe mister smarty pants oughta be sellin' dictionaries instead of vacuums!)* 5. hocus-pocus; card tricks; sleight of hand 6. spontaneous made-up language *(eg. "What do you mean, Brzzzzttyyyiffft pyowzchks&'? What's with all the gobbledygook?" she asked.)* 7. speaking in tongues

grid-second
-*noun* A unit of time measurement which is somehow connected to some sort of harmonic energy grid, documented by a UFO researcher named Bruce Cathie, that runs everywhere all over the Earth, at specific intervals somehow involving the number 144. Refer to Appendix 144.

assemblage point
-*noun* In the writings of Carlos Castaneda, the assemblage point is a sort of energy plexus located on the luminous egg, behind and betwixt the shoulder blades. Moving the assemblage point through some sort of shamanic act of will shifts one's entire reality to a new dimension, allowing one to see things from an altered perspective.

betwixt
-*prep.* Between.

betwixt and between
-*colloq.* Between and betwixt.

brainthrill
-*noun* My latest word-invention. It means "the state or action of being happy in the brain." Brainthrill is often the result of intellectual stimulation, or some other such "Eureka" moment where a figurative light bulb switches on above the head. Other types of thrill might include *heartthrill*, *soulthrill* or *crotchthrill*.

brainthriller
-*noun* That which causes or instigates a brainthrill. *(eg. The Orbital Gnomon Star Power Chord is a real brainthriller!)* Other types of thriller might include the *soulthriller*, the *heartthriller* or the *crotchthriller*.

Eureka!
-*noun* A verbal expression immediately following a moment of delighted enlightenment. Etymlogy lesson: Apparently, so the story goes, an engineer named Archimedes discovered that an object's volume could be measured by immersing it in water and measuring the volume of displacement of the surrounding water (or something like that). When he made his discovery, he jumped out of the bath and ran naked through the streets yelling, "Eureka! Eureka!" The story ends with Archimedes getting killed by a Roman soldier while contemplating calculations in the sand.

collective memory
-*noun* Have you ever noticed that "learning something new" feels the same as "remembering"? This seems to support the idea of a collective consciousness. Not only consciousness, but also group *memory*. Learning something new is really regaining access to the ever-present body of universal knowledge that sits patiently waiting.

vacuum
-*noun* A really, really empty place. Imagine Darth Vader's heart, when he hasn't eaten his green veggies. Which raises an interesting point: light travels through a vacuum quite easily, whereas sound does not. This is one instance where the analogy between light and sound seems to fail.

dog's breakfast
-*noun* A chaotic mess, hopeless situation, botched attempt or missed target.

thingamajigger
-*noun* Whatchamacallit.

ratio
-*noun* The mathematical division of two numbers. "How many times bigger is this thing than the other thing?" Ideally, a ratio is expressed as a fraction such as 8/5, or with a colon in between the numbers like 2:3. The use of the colon in this instance is reminiscent of analogies such as "bird is to flock as wolf is to pack," which could be expressed more simply as the proportion:

wolf : pack = bird : flock

Sometimes the "=" is replaced by a "::", like so:

crows : murder :: geese : gaggle

or

goose : good :: gander : good

interval
-*noun* The distance between any two musical notes; in other terms, the ratio between two frequencies. In the twelve-tone Chromatic Scale, this distance is typically expressed in some number of semitones. Specific names have been attributed to these intervallic distances.

scale
-*noun* A collection of musical intervals; extremely similar to a palette of coloured pigments or a collection of planets circling a star.

N.T.S.
-*abbr.* Not to scale.

spectralism
-*noun* A cool word that I wish I'd invented. It refers to a style of musical thought where scales are transcended and sound is perceived and manipulated as a continuous and open-ended spectrum - just like *colour!*

microtonal music
-*noun* The easiest way to compose microtonal music that I can

SECTION VII: GLOSSANDO 177

think of is by asking a novice musician to play the fretless bass. A trombone, slide guitar, or singing saw would also do the trick.

novice
  -*noun* No vices. Innocent. Ignorant. Clean slate. Empty vessel. Pure potential.

planet
  -*noun* A globe in space, vast and shining like in the movies.

year
  -*noun* The space betwixt birthdays; one cycle of a planet's sine wave.

ripitition
  -*noun* The repeating of stuff. I propose that this new spelling replace the archaic spelling on the grounds that the high frequency of the letter *i* truly reflects the meaning of the word.

symmetry
  -*noun* See *symetry*.

symetry
  -*noun* Anagram of *mystery*. (eg. How could there possibly be so much symetry in the universe?)

multiverse
  -*noun* What I really mean when I say "universe." After all, if "as above, so below," then:

  quark : atom :: our_universe : some_sort_of_multiverse

oroborus
  -*noun* A snake that swallows its own tail.

moebius strip
  -*noun* 1. Imagine, if you will, that you are traveling down a long highway. To your surprise, you find that the road never ends. Stranger still, you notice that the scenery seems to repeat itself. You seem to be stuck in some sort of infinite loop. This loop, however, is unlike any broken record player you've ever experienced: each time you revisit a piece of scenery, you're on the *flipside* of the asphalt, observing the roots of the trees instead of the branches, then the branches, then the roots,

then the branches, then the roots, then the branches, then the roots, then the branches, then the roots, then the branches, then the roots.... This is what we call the moebius strip. Don't try this one at home. 2. "Just because you're paranoid doesn't mean they're not watching you."

maxim
See *axiom*.

interrobang
 -*noun* A cross between a question mark and an exclamation point. Punctuation indicating surprise, disgust, confusion, disagreement, falling off a pyramid of cubes (almost - @!#?@!), etcetera. According to history it was invented by an advertising guru named Martin Speckter in New York in 1962. It was actually added to a few typewriter keyboards in the 60's but it just never caught on. I wish it would.

Fig. 4: Interrobang

history
 -*noun* Whose?

palindrome
 -*noun* Anything that spells the same backwards as forward. *(eg. 2002, 1001001, Bob, Abba, Boob, Boooooooob, Mam, Gag, Toot-toot, I'm A Goddam Mad Dog Am I?)*

SECTION VII: GLOSSANDO

gamut
 -*colloq.* The whole range. *(eg. The guitarist spans the gamut from blues to classical to jazz to funk and everything inbetwixt.)* origin: from Greek *gamma* and Latin *ut* which was the name of the lowest note in the original *Solfège* scale of Ut Re Mi Fa Sol La before it was de-harmonized into Do Re Mi Fa Sol La Ti.

Solfeggio
 -*noun* An ancient musical scale containing six tones which are reputed to possess divine mathematical and spiritual resonance. The tones are called Ut, Re, Mi, Fa, Sol, and La and their frequencies are arranged in a simple structure, like so:

La	852 Hz
Sol	741 Hz
Fa	639 Hz
Mi	528 Hz
Re	417 Hz
Ut	396 Hz

frequency
 -*noun* The number of cycles per unit of time in any situation involving ripitition (waves). *(eg. The sound frequency of 528 cycles-per-second, also known as "Mi" in the Solfeggio scale, is known to repair DNA.)*

hertz
 -*noun* Cycles-per-second, a measure of frequency. Abbr. Hz. *(eg. When Martha sang the frequency of 741 Hertz, she began to levitate. When she modulated her pitch to 852 Hertz, the lights dimmed and the furniture began to spin like whirling dervishes.)*

modulation
 -*noun* Any act of change. *(eg. The only guarantee in life is... modulation.)*

wavelength
 -*noun* The distance between crests of a wave. The wavelength is also the inverse of frequency - therefore, a higher frequency has a short wavelength and a long wavelength will produce a lower tone. In terms of light, for example, red is a lower frequency and has a longer wavelength than blue.

Just Temperament
  -*noun* A system of tuning calculation which favours the use of small-numbered ratios. Just Temperament is more natural but less versatile than Equal Temperament.

Equal Temperament
  -*noun* A system of tuning calculation in which all twelve subdivisions of the octave are equally spaced, making transposition possible without re-tuning. This system has its sacrifice - many of the intervals are "out of tune" with nature.

octave
  -*noun* The doubling or halving of a sound frequency. One complete cycle of a typical scale. Twelve semitones. Twelve frets on the guitar. The first overtone. The most inert of all musical intervals - a C is a C is a C is a C is a C is a C is a C is a C is a C.

temperature
  -*noun* The speed of moving molecules in a substance; in other words, "how hot" is really "how fast." So, when I'm speeding out on caffeine, I might actually be overheating my brain. Yikes! Some like it hot.

margin of error
  -*noun* How out of tune something is allowed to be before it is considered to be something other than itself. For example, when was the exact time of the full moon? Yesterday, or this morning, or this very second? Oops - missed it! Or, perhaps its *effects* are really the determining factor, and therefore should be considered "full" for about a two- or three-day period. Whatever suits your purposes, I suppose. See Appendix ± for an in-depth technical analysis.

babooshka
  -*noun* Also known as babushka, matreshki, stacking dolls, etcetera. Now also a popular casino game, these dolls nestle one inside the next like layers of an onion.

subdominant minor
  -*noun* 1. The musical interval of a minor sixth. A minor third above the subdominant. In medieval music theory, it is called the *tetratone,* equal to four whole tones. This authour (me)

SECTION VII: GLOSSANDO   181

calls it *the golden interval* because Phi is 1.618 and the just minor sixth is 1.600. 2. A BDSM fetishist who digs for tetrahedrons.

, eh?
–colloq. A colloquialism of Canadian origin, popularized in the '80s by a couple of drunken comedy heroes, which is tacked on at the end of a sentence and which expresses a sentiment similar to the question, "Do you know what I mean?" or, "Don't you agree?"

the '80s
-noun, plural The decade starting in MCMLXXX.
(eg. ...so I sez to me pal, "The '80s were f*&^in' great, eh?")

snake oil
-noun A medical cure or antidote, or any supposed self-improvement scheme, which has more value as entertainment or placebo than anything else; typically sold from the back of a van with blacked-out windows or a chuckwagon equipped with secret compartments.

"you shoulda been there!"
-colloq. Sadly... no matter what you're doing, you're missing out on everything else. One feasible antidote would be to learn the skill of occupying multiple locations simultaneously, thereby avoiding any possibility of absenteeism.

"the grass is always greener on the other side of the fence"
-colloq. A poem, perchance, to hint at the notion:

every blade
grateful as the next
humbly drinking the sun
in all their greenliness
and singing together the eternal question
"what is this fence they speak of?"

tone-deaf
-adjective The practically fictitional inability to distinguish musical pitch, exemplified by the axiom "I couldn't carry a tune if I had a bucket to carry it in!" Sometimes the bucket is just not the right size or shape of a bucket. Everybody's innate bucket

is very special - much like a beautiful snowflake - and uniquely tuned to some quirky collection of musical data configurations. Buckets are expandable, re-shapeable and *learnable*. Buckets are inventable. Buckets can collect like a good wardrobe or a bunch of dead bugs on pins behind glass. Or, a bunch of bugs at the bottom of a window, not on pins. Or, how about that mysterious mildew crap that keeps crawling out from another dimension and into the cracks of the bathroom sink and shower... *weird!* Alas, I digress. And, all things considered (such as the hugeness of eternity compared to the smallness of everything we can ever know in only these 3-ish dimensions), if a person can simply experience the beauty of enjoying a piece of music, then they are already demonstrating a miraculous level of vibrational perception - it can only get better from there by singing and learning to play music! Merely listening closer or with a new perspective can create magical spirals of inspiration within a being. Sound literally bends reality.

dance
-noun ...yet another dimension. See also *theatre*.

inspired
-adjective In spirit.

pinna
-noun The bit that Van Gogh was done with. Observe the serendipitous similarity to the treble clef:

times-tables
-noun Time tends to turn the tables. And turn and turn.

overtone series
-noun A collection of tones which automatically appears over and above a single fundamental tone. Also known as harmonics, the overtone frequencies follow a simple pattern of ratios: 2/1, 3/1, 4/1, 5/1, 6/1, 7/1, etc. This is equivalent to dividing a vibrating string (ie. skipping rope) into 2 equal pieces to create an S-shaped rotation, 3 pieces into a revolving "W", 4 pieces, 5 pieces, etcetera. A musical instrument's overtone mixture defines its tonal quality to a large extent. Various factors affect the mixture of overtones, such as the shape and volume and mass

SECTION VII: GLOSSANDO 183

of the instrument, the thickness or tightness of the strings, how softly it's being played, how new the strings are (and the condition of various other components), how sweaty the player's hands are, what sort of electronic effects are being applied to the signal, the distance of the instrument to the listener's ears, and whether or not tetrahedral carbon allotropes have been embedded into the varnish of the instrument. Oddly enough, a tube which is closed at one end - such as blowing across a half-empty (or half-full, attitude depending) beer bottle - only produces odd-numbered overtones. Harmonics occur naturally, but they can also be induced via various performance techniques. See also *tetractys*.

harmonograph
-*noun* A table with two slots cut in it, through which two separate penduli undulate with their periods of culmination tuned to various musical intervals. One pendulum has a pen attached to the top of it, and the other moves a flat drawing surface; a loopy-looking image is produced.
  exempli grata:

FIG. 4: PERFECT FIFTH (3:2) - NOTICE THE NUMBER OF BUMPS ON EACH SIDE

twelfth
-*adjective* 1. notice there's only one vowel out of seven letters! (see Appendix 142857) 2. twelfth = tw_elf_th = th_tw_elf = the_twin_elf; therefore, sixth = one_elf.

tetractys
 -noun A triangular arrangement of dots which illustrates the overtone series, among other things. The typical classic tetractys only goes up to four rows, but the following extended version offers more detailed harmonic insight:

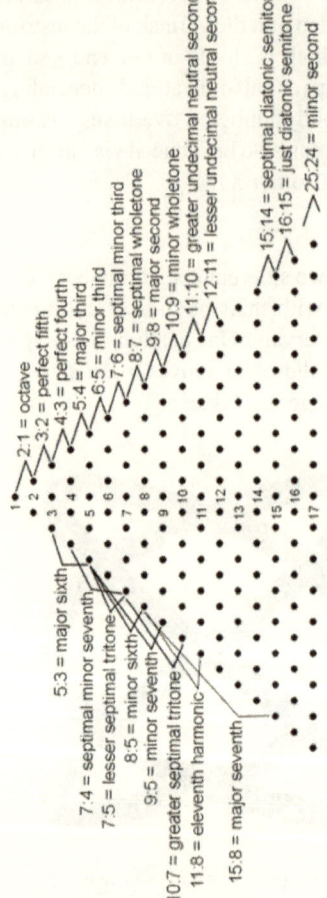

Fig. 4: Extended Tetractys

tetrachromat
 -noun An organism whose visual equipment uses four colour sensors. Some birds and marsupials are tetrachromats, with a sensor tuned to the ultraviolet range; the typical human has only three such sensors (known as cones) but supposedly some females are born with a fourth sensor which is tuned to yellow.

Section VIII

BIBLIOGRAPHY

*If you steal from one another its' plagiarism;
if you steal from many its' research."*
                            -Wilson Mizner

PRINT SOURCES:

Antara Amaa-Ra, Solara. *11:11*. Charlottesville: Star-Borne Unlimited. 1992.

Ashton, Anthony. *Harmonograph: A Visual Guide To The Mathematics Of Music*. Wales: Wooden Books. 2003.

Brandreth, Gyles. *The Joy of Lex: How to Have Fun with 860,341,500 Words*. New York: William Morrow and Company, Inc. 1980.

Bugler, Caroline. *The Complete Handbook Of Astrology*. London: Marshall Cavendish Books. 1992.

Cavendish, Richard. *Mythology: An Illustrated Encyclopedia*. London: Orbis Publishing, Ltd. 1980.

Clark, Rosemary. *The Sacred Tradition In Ancient Egypt: The Esoteric Wisdom Revealed*. St. Paul: Llewellyn Publications. 2000.

Cytowic, Richard E. *The Man Who Tasted Shapes: A Bizarre Medical Mystery Offers Revolutionary Insight Into Emotions, Reasoning, And Consciousness*. New York: Putnam. 1993.

De Bono, Edward. *Lateral Thinking: A Textbook Of Creativity*. London: Penguin Books. 1990.

Freke, Timothy and Gandy, Peter. *The Hermetica: The Lost Wisdom Of The Pharaohs*. New York: Tarcher/Putnam. 1999.

Gadalla, Moustafa. *Egyptian Cosmology: The Absolute Harmony*. Erie: Bastet Publishing. 1997.

Graham, Robert. *Night Vision: The Powers Of Darkness*. Hilmarton: Matrix. 2000.

Hofstadter, Douglas R. *Godel, Escher, Bach: An Eternal Golden Braid*. New York: Basic Books. 1979.

Horowitz, Leonard G. and Barber, Joseph. *Healing Codes For The Biological Apocalypse.* Sandpoint: Tetrahedron Publishing Group. 1999.

Livio, Mario. *The Golden Ratio.* New York: Broadway Books. 2002.

Martineau, John. *A Little Book Of Coincidence.* Wales: Wooden Books. 2001.

Parramón, José M. *Color Theory.* New York: Watson-Guptill Publications. 1989.

Peter, Dr. Laurence. *Quotations For Our Time.* London: Black Cat. 1979.

Ravenwolf, Silver. *Solitary Witch.* St. Paul: Llewellyn. 2003.

Robbins, Tom. *Skinny Legs And All.* New York: Bantam Books. 1991.

Salem, Lionel and Testard, Frédéric and Salem, Coralie. *The Most Beautiful Mathematical Formulas.* New York: John Wiley & Sons, Inc. 1992.

Sitchin, Zecharia. *The 12th Planet.* New York: Avon Books. 1976.

Sobel, Dava. *The Planets.* New York: Penguin Books. 2006.

Spiekerman, Erik and Ginger, E.M. *Stop Stealing Sheep & Find Out How Type Works.* Mountain View: Adobe Press. 1993.

Stewart, R.J. *Music Power Harmony: A Workbook Of Music & Inner Forces.* London: Blandford. 1990.

Stillwell, John. *Mathematics and Its History.* New York: Springer. 2002

Toth, Max and Nielsen, Greg. *Pyramid Power.* New York: Destiny. 1976.

Tufte, Edward R. *Envisioning Information.* Connecticut: Graphics Press. 1990.

West, John Anthony. *Serpent In The Sky: The High Wisdom Of Ancient Egypt*. Illinois: Quest Books. 1993.

Yatri. *Unknown Man*. New York: Simon and Schuster. 1988.

Zukav, Gary. *The Dancing Wu Li Masters*. New York: Bantam. 1980.

INTERNET SOURCES:

http://a-440.net/pitch.htm#cents

http://www.bazookajoe.com/

http://crystalinks.com/sound_frequencies.html

http://dictionary.reference.com/search?q=moire

http://dsc.discovery.com/news/2007/03/27/hexagon_spa.html?category=space

http://en.wikipedia.org/wiki/Phi_%28letter%29

http://home.c2i.net/greaker/comenius/9899/pythagoras/pythagoras.html

http://home.swipnet.se/freakguitar/scales.html

http://hometown.aol.com/codeufo/eltanin.html

http://hypatia-lovers.com/geometry/Divine_Proportion.html

http://news.nationalgeographic.com/news/2002/07/0710_020710_chadskull.html

http://phycomp.technion.ac.il/~nika/diamond_structure.html

http://tonalsoft.com/

http://violet.pha.jhu.edu/%7Ewpb/spectroscopy/basics.html

http://wingedeyesymbol.homestead.com/

http://www.ac.wwu.edu/~stephan/webstuff/ratio.digits.html

http://www.aloha.net/~hawmtn/horus.htm

http://www.andrews.edu/~calkins/math/biograph/biopytha.htm

http://www.bam.ie/bambrat/light1.htm

http://www.blazelabs.com/f-p-solids.asp

http://www.businessballs.com/clichesorigins.htm

http://www.colorado.edu/greeks/alphabeta.html

http://www.cseligman.com/text/sky/rotationvsday.htm

http://www.decoz.com/pythagoras.htm

http://www.entropysite.com/students_approach.html

http://www.etymonline.com/
    index.php?search=moire&searchmode=none

http://www.geocities.com/timessquare/alley/1557/fonts1.htm

http://www.greatdreams.com/144.htm

http://www.greatdreams.com/gem1.htm

http://www.greece.org/samians/pythagoras.htm

http://www.innerpotential.org/pages/article/maldek.html

http://www.innerpotential.org/pages/article/sun.html

http://www.keyway.ca/htm2002/greekal.htm

http://www.m-base.com/cnmat_ucb/Symmetry_Movement.html

http://www.medieval.org/emfaq/harmony/pyth.html#intro

http://www.mjh434.net/tma02/biblio.htm

SECTION VIII: BIBLIOGRAPHY

http://www.phy.mtu.edu/%7Esuits/scales.html

http://www.quinion.com/words/weirdwords/ww-int1.htm

http://www.rdrop.com/users/tblackb/music/temperament/stoess.htm

http://www.space.com/astronotes/astronotes.html

http://www.space.com/scienceastronomy/solarsystem/diamonds_planets_990930.html

http://www.suisyounoentaku.com/13moon_share/28med22.htm

http://www.tonalsoft.com/enc/g/gamut.aspx

http://www.tortuga.com

http://www.treeoflifeschool.com/Newsletters/happy-5765.htm

http://www-groups.dcs.st-andrews.ac.uk/~history/Mathematicians/Pythagoras.html

VIDEO SOURCES:

Amateau, Rod. *The Garbage Pail Kids Movie.* Produced and directed by Rod Amateau. 97 min. MGM DVD, 1987. DVD.

Kawamoto, Toshihiro. *Cowboy Bebop: The Movie.* Written by Keiko Nobumoto. 115 min. Sony Pictures, 2001. DVD.

Section IX

ACKNOWLEDGMENTS

No-one can exist in a vacuum!

Acknowledgments

Thanks to family & friends for their support & inspiration - during this project and in general - including but not limited to the following people: Anthony Metivier & Rob Read at Produce Section Press, Mum & Dad, Tanya & Brian Logan, D. Ian Smith, Chris Bose, Shane Smith for the introduction to "Healing Codes", Sandra Bandura, Carolyn Hines, Linda Rightmire, Chuck St. John, J.R. Adam, Christina Meadows, Mike Schubert for help with audio generation, Stephanie Stephens and the amazing two-tone tuning fork, Gary Faustman, Paul Liddy, the Bartsch's, Stu MacKay-Smith, Darlene Fair, Andrée Beauchemin, Adam MacKay-Smith, Iain McLaren, Dana Shoesmith, Clint Neighbor, Rachel Vyse, Ryan MacDuff, Jon Treichel, Travis Wright, the Nukina's, Barbara Zimonick, everyone at B.C. Festival Of The Arts' Electronic Music Producers Seminar, Steve Coleman (whose theories springboarded my discovery of "Modal Mirrors"), Donovan King Pettigrew, Claudette Laffey, Vaughn Warren, Kiera Merriam, Marcella Huberdeau, Laurie Payne & his Golden Mean expressions, Steven Hurst, James P. Bethell, Grant Hartley, Dave Burgess and The Grind, the jam scene, Jacquie Brand, Julia Appley-Mitra, The Community Arts Council of Kamloops, Karmin Poirier, Alex Muendel, Graham Lazarovich, Fuse Magazine, Wiley, Ken Lawson, Brad Harder, Bobbi Mitchell, Arizona Dave, Leslie Charman, Janet Michael, David Ross, WCTC, Melissa Thomas, Project X Theatre, Steve Marlow & Brant Zwicker at The X 92.5FM, the Badgers, all my students, the miscellaneous stranger at Denny's who unknowingly contributed the pretzel joke, the Library & my various other favourite writing/research/hangout spots, and all the jammers, musicians, DJ's, artists & other fine humans & animals who've bounced ideas back and forth with me...

Sean Alexander Luciw

The fat lady sings and the lights go out.
The crowd goes wild!

www.ingramcontent.com/pod-product-compliance
Lightning Source LLC
Chambersburg PA
CBHW020927090426
42736CB00010B/1063